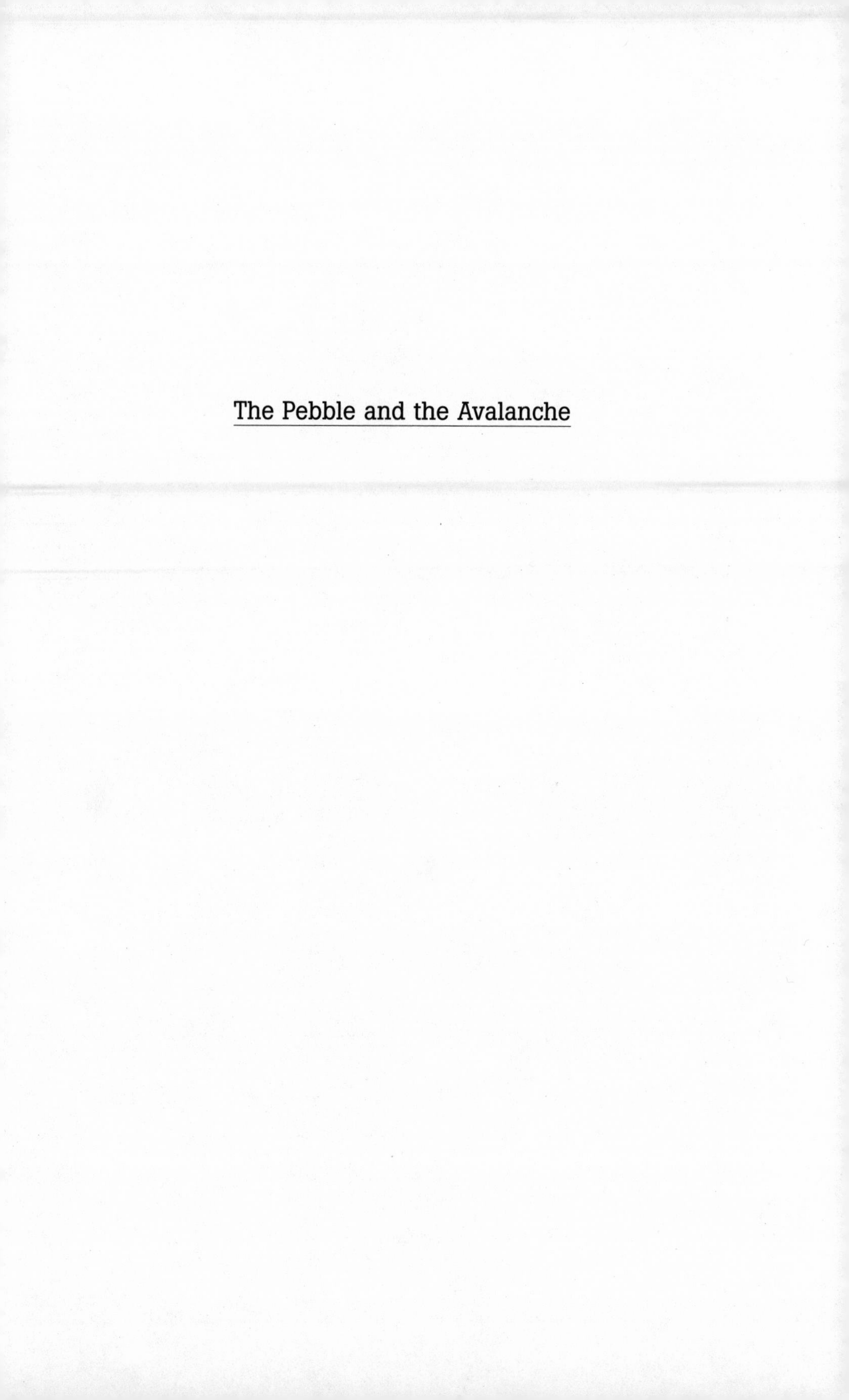

The Pebble and the Avalanche

The Pebble and the Avalanche

How Taking Things Apart Creates Revolutions

Moshe Yudkowsky

BK

Berrett-Koehler Publishers, Inc.
San Francisco

Berrett-Koehler Publishers, Inc.
235 Montgomery Street, Suite 650
San Francisco, CA 94104-2916
Tel: (415) 288-0260   Fax: (415) 362-2512   www.bkconnection.com

Ordering Information
Quantity sales. Special discounts are available on quantity purchases by corporations, associations, and others. For details, contact the "Special Sales Department" at the Berrett-Koehler address above.
Individual sales. Berrett-Koehler publications are available through most bookstores. They can also be ordered directly from Berrett-Koehler: Tel: (800) 929-2929; Fax: (802) 864-7626; www.bkconnection.com
Orders for college textbook/course adoption use. Please contact Berrett-Koehler: Tel: (800) 929-2929; Fax: (802) 864-7626.
Orders by U.S. trade bookstores and wholesalers. Please contact Publishers Group West, 1700 Fourth Street, Berkeley, CA 94710. Tel: (510) 528-1444; Fax (510) 528-3444.

Printed in the United States of America

Berrett-Koehler books are printed on long-lasting acid-free paper. When it is available, we choose paper that has been manufactured by environmentally responsible processes. These may include using trees grown in sustainable forests, incorporating recycled paper, minimizing chlorine in bleaching, or recycling the energy produced at the paper mill.

Library of Congress Cataloging-in-Publication Data

Yudkowsky, Moshe, 1956–
The pebble and the avalanche : how taking things apart creates revolutions / Moshe Yudkowsky.
p. cm.
Includes bibliographical references and index.
ISBN-10: 1-57675-294-1; ISBN-13: 978-1-57675-294-4
1. Technological innovation. 2. Diffusion of innovations. I. Title.

First Edition
10 09 08 07 06 05   10 9 8 7 6 5 4 3 2 1

Cover by Crowfoot Design
Interior Design by Dianne Platner
Production by Publication Services, Inc.

To my wife, Rachel, whose love, kindness, patience, and support made this book possible:

"Many women have accomplished valor,
But you outshine them all."
Proverbs 31

To my children, Eliezer and Channah, for their love, help, and support over the years:

"He has blessed your children among you."
Psalms 147

In memory of my son, Yehuda Nattan Yudkowsky:

"Who loved his fellow man, and brought them close to the Torah."
Ethics of the Fathers, 1:13

Contents

Part II. Case Studies: Two Hundred Years of Revolutions

Part III. Business Strategies: How to Cope, How to Fail, and How to Predict the Future

Preface

This book explains how to understand, create, and apply revolutions in business and technology.

A few years ago, my colleagues and I met at the Boston airport to discuss the future of our organization, an industry group in the field of telecommunications. We'd hired a professional moderator to lead the discussions, and at one point the moderator had us working on the significant inventions of the past thirty years, the innovations that had transformed telecommunications. We generated a timeline with some interesting items on it (you'll see some of them in Chapter 1), and then the moderator asked a rather ho-hum, standard question just to move the conversation along: "What do these inventions have in common?"

The answer hit me like a flash of lighting: Each of the important inventions and revolutions in our field started, just as the subtitle of this book says, when people took things apart. The key innovations really had something in common—something exciting and unexpected. I jumped up and explained my idea to the group. Everyone nodded, but one person did nothing to show he'd even heard me. This person is a creative and highly competent individual, a technical person like me. I was concerned about his reaction; if he didn't agree, then clearly I was missing something. Yet he just sat there, staring at the whiteboard without saying a thing.

"Don't you agree with what I'm saying?" I asked.

"I'm figuring out how to use your idea to make money," he replied.

That's when I knew I was onto something important.

This book provides ideas, methods, and examples and shows you how to use them to create useful and exciting innovations. "Taking things apart creates revolutions" is the simple, one-line answer, but the more I looked into this relatively simple idea the more rich, the more interesting, and the more fun it became. I hope you'll enjoy the rest of this book as I guide you through the details.

The main focus of the book is on business and technology. The ideas in this book apply across a wide range of activities; I've included discussions about government and economics, but have left out any mention of medicine, health care, or religion.

Part I of the book presents the fundamental ideas. This section discusses how taking things apart works, how to categorize the different ways of taking things apart, and some of the implications. Also included is information about the benefits to expect—the payoff that makes all the hard work worthwhile.

Part II of the book consists of case studies. Three revolutions in technology—dating from the nineteenth, twentieth, and twenty-first centuries—provide examples of how these revolutions work in detail.

Part III provides strategies for how to cope with revolutions. There's discussion on how to avoid being buried by an "avalanche" and warnings about strategies that simply don't work, such as running in front of the onrushing avalanche and yelling at everyone to stop. There's also a chapter about some up-and-coming revolutions—a few places where it's possible to see the first "pebbles" that signal the oncoming avalanche.

Finally, a few things about what this book does *not* do. "Taking things apart" does create revolutions, but then again there are many revolutions that don't fit this mold. The book doesn't claim that all innovation, all technological revolutions, and all changes in human society are explained by this one theory. What I'm discussing is important, but not all-inclusive. There are at least three broad categories of revolutions.

One category is "replacement" revolutions, which start with the introduction of a replacement technology. An example of this is the steam engine, which replaced and/or supplemented the existing power sources: muscle, wind, and water. And then there are revolutions that start when someone invents completely "new physics" and introduces capabilities that simply weren't there before—radios, X-rays, nuclear power plants, and radiation therapy. Both the "replacement" revolutions and the "new physics" revolutions are relatively scarce because they rely on scientific breakthroughs, and science doesn't produce breakthroughs on a regular schedule.

But this book is about a third category of revolutions, revolutions that are far easier to create, revolutions that account for much of the progress we've seen in the past thirty years. These revolutions are based on taking things apart, a process I call *disaggregation*. . . .

Moshe Yudkowsky
Chicago, Illinois, USA

Acknowledgments

The staff at Berrett-Koehler, the publishers of this book, deserve a special thanks; they're a wonderful team of people who look on authors as partners, and they made me feel very welcome in a personal way. (All of my author friends at other publishers are jealous of the treatment Berrett-Koehler gives its authors.) Jeevan Sivasubramaniam shepherded this book through to production, and he was the first person to spot its potential. I owe a special debt of gratitude to Steve Piersanti, editor and publisher, who helped me find my voice. Steve and Jeevan provided invaluable advice and support, and the book simply would not have been possible without them.

Berrett-Koehler authors meet on a regular basis to discuss their work, to exchange ideas, and provide suggestions. I'd like to thank them for their support. Dick Axelrod and Mark Levy, in particular, answered urgent phone calls and explained some of the finer points of writing a business book.

My colleagues in the technical community provided encouragement. John Kelly of SpeechTEK and Bill Meisel of TMA Associates gave me the opportunity to speak at their respective conferences about the ideas in this book just as I started my consulting practice, which helped me refine my thoughts and get feedback from the technical community. Judith Markowitz, author and consultant, provided advice, tea, and muffins on a regular basis.

The members of General Technics—a loose-knit organization of friends who can be found literally anywhere from Antarctica to the Arctic Circle, doing jobs from ocean research to flying probes in outer space—provided answers, support, and encouragement.

I would also like to surprise two people. In reply to a comment, friend and science fiction author Phyllis Eisenstein made a pointed remark to me years ago: "You can write if you want to, Moshe." I took her advice, and here I am today. My thanks to author, consultant, and friend Bruce Schneier, who years ago encouraged me to write a business book; I've finally taken his advice. My thanks go to Bruce for reading a draft of this book with pen in hand.

I would also like to thank my friends, my colleagues, and the members of the Jewish community for their assistance and support during difficult times.

According to Jewish tradition "the last is the most beloved." I would like to thank my wife, Professor Rachel Yudkowsky, M.D., for her support over the years and especially during the transition from "salaryman" to consultant and author. She read multiple drafts of each chapter in this book—if that's not true love, I don't know what is.

Even though I am thankful to the people who read and commented on drafts, any errors or omissions in this book remain my responsibility (with a background in technology, I know that they must be there; I just don't know where they are). I welcome feedback and the opportunity to make clarifications. You can send comments through the book's Web site, at http://www.PebbleAndAvalanche.com.

Disaggregation: Why the Sum of the Parts Is Greater Than the Whole

Part I

Part I explains the mechanics of how to take things apart in order to start revolutions in business and technology.

What things to take apart

Benefits you should expect

How to assess an innovation for revolutionary potential

A step-by-step approach to apply these ideas to your particular problem

Chapter One

Disaggregation: The Driving Force of Revolution

One of the first safety rules I learned while hiking in the mountains was to never toss pebbles or stones down the side of a mountain. There's the danger of hitting someone—a pebble that falls a thousand feet can do an impressive amount of damage. The other danger is starting an avalanche. It's a tiny little pebble, true; a pebble that size can dislodge only another few pebbles, true; but if enough pebbles start to tumble, soon the large rocks start to move, and your one little pebble triggers a landslide.

An avalanche releases energy—a really impressive amount of energy. Shift a few pebbles, *take apart* the structure that's holding the rock formations together, and suddenly you release an incredible, unstoppable force that transforms the landscape. Avalanches snap trees in half, shove boulders out of the way, and cut a huge swath out of forests. However, despite their massive power, when avalanches stop, you've still got the most of the pieces you started out with. All the pebbles that started off at the top of the mountain fall to the bottom—the pebbles aren't gone, they're just arranged differently—and now you have a nice collection of interesting pebbles, conveniently located here at the bottom of the mountain. They can be cut, polished, and made into jewelry; they can be used to build walls and pave garden paths. They're still useful in many ways, and so is all the other debris that's been brought down by the avalanche.

This book is about how this same idea applies to everyday life—technology and the business of technology in particular. Technology and the business of technology have structure; if you *take apart* that structure you can unleash an avalanche that has tremendous energy, one that can change the entire landscape. Avalanches smash old businesses into smithereens; sometimes the businesses vanish entirely, and sometimes a few pieces survive. Old, comfortable business and technical relationships snap under the stress of the avalanche. New opportunities appear in the empty spaces left behind.

Not only that, but when the avalanche is over, you've still got, for the most part, the pieces of technology you started with in the first place. And just like the pebbles in an avalanche, these pieces of technology can be improved, used over again, and made infinitely better than before. These pebbles are no longer jammed together in some lump that's impossible to use, and they're not hidden under larger rocks where it's impossible to get at them. They're accessible, lying around waiting to be picked up. A small pebble is much easier to handle than a large rock. Pebbles are easier to polish, cut, decorate, and fit into a beautiful mosaic.

What are the pebbles that make up the structure of technology? They include the usual things we think of: nuts and bolts, electronics, manufacturing plants, and chemicals to name a few. Software, processes, and work flows are just as important to technology, even though they're not tangible. Other equally important pebbles—ones that can also be taken apart to start an avalanche—are the social pieces of the technological landscape. These pieces include government regulations, business ideas, intellectual property law, patent rules, and dozens of other social structures that govern the business of technology, how the pebbles can be used, and what structures can be built in the first place.

Taking Things Apart: Recent Revolutions

Here's a list of some revolutionary changes that happened over the past thirty years. Each one unleashed an avalanche and completely changed both technology and the business of technology.

- The Internet, which provided obscure services that later became quite popular, such as e-mail and Web browsing.
- AT&T's divesture into separate long-distance and local phone companies, which ultimately drove the price of long-distance service to near zero.
- Personal computers brought the benefits of computing to everyone, not just the lucky few.
- The World Wide Web transformed how we share information.
- Open source software provides excellent—and often free—software and challenges the entire software industry to compete and improve.
- Telephone calls over the Internet are about to make classical telephone systems utterly obsolete.

The items on this list don't seem to have much in common, do they? In fact, *every single one* of these important revolutions started by taking something apart:

1. The Internet? Before the widespread use of the Internet, each manufacturer had its own idea of how to transfer data between computers. This made it difficult, or sometimes practically impossible, to create networks or send interesting information—e-mail, music files—between computers from different manufacturers. The Internet describes a common set of methods to transfer data between computers, which broke an entire piece of technology out from under the manufacturers' control and made it possible to send data between any two computers on the Internet.
2. AT&T? Well, that's easy: the company broke apart into separate entities, a change in the business of technology that had profound implications for the technology itself. Competition has

since driven the price of long-distance calls to be next to nothing, and new services are everywhere.

3. Personal computers? The parts that go into computers stopped being custom-made for each different model; instead, the relationship between computer model and, say, the disk drive was broken and all disk drives became commodity parts. Computers now run standard operating systems—the operating system is no longer part and parcel of the hardware, as it was in the old days.
4. World Wide Web? Prior to the development of the Web, you couldn't just send someone an electronic document and expect them to be able to read it. When a supplier sent me a catalog on a disk, I had to install a special "viewer" program to read the catalog—and each and every supplier had its own program! There was a strong connection between the document and the viewer program. With the Web, electronic documents can be viewed using any Web browser because the Web breaks the connection between a document and some particular brand of viewer software.
5. Open source software? The functions of writing, documenting, selling, maintaining, and improving software can be performed by separate groups instead of by the more traditional single company.
6. Telephone calls over the Internet? Instead of paying the phone company to route your call *and* provide you with the bandwidth to carry your voice, Internet telephony lets you separate the two—you can use any Internet connection to carry the call.

There are more examples, from the past thirty years, the past two hundred years, and the past two thousand years; we'll cover some of them later in this book. The basic idea remains: many important revolutions in technology started by taking things apart. These revolutions act as avalanches, as tremendous forces that sweep aside the old, clear the way for the new, and leave interesting and important pieces behind.

Disaggregation: The Science of Taking Things Apart

"Taking things apart" is a nice expression, but it doesn't really give the exact flavor of the process, so I use the term *disaggregation* to explain the idea. Here's the explanation.

Often a particular technology, or a business of technology, seems to be made up of one solid piece; but if we inspect it carefully we can see that it's really made out of individual pieces—pieces we can take apart if we're careful. The business of telephone service is an excellent example. In the United States until 1984, and aside from a few small competitors, AT&T did everything: it provided long-distance service and local service, installed and owned the wires on the city streets, installed the wires in homes and businesses, and even installed and owned the telephones in people's homes. There was no technical reason why it had to be that way—the electric company didn't insist on owning and installing everyone's electrical appliances!—but through social and legal conventions, all these functions were aggregated, that is, stuck together, into one entity.

How do I describe taking AT&T apart? I don't want to use words like *destroyed* or *smashed* because it was a very careful processes. Nothing was destroyed; all the equipment remained in place—in some buildings, the workers painted white lines on the floor to show that AT&T's equipment was "here" while the local telephone company's equipment was "there." Telephone service wasn't disrupted—calls continued without any interruption. The best way to describe taking apart AT&T is to look at the results. The various pieces of phone service are no longer aggregated into one solid mass—they are *disaggregated* into separate entities. The pieces are no longer bound together with the social glue. *Disaggregate* is the word I'll use throughout this book to describe this process of taking things apart, of breaking connections, and of dismantling the infrastructure of technology and its businesses.

Did disaggregating AT&T make phone service better? Absolutely: top to bottom, left, right, and sideways. The first advances came in long-distance service: the quality of connections improved at the same time prices dropped. Then came new services that AT&T would never have offered, like prepaid phone cards. And look at the different styles of telephones! In the old days at AT&T, all telephones were black; today they come in every conceivable shape and color, and a few that are outright ridiculous. These benefits are typical of disaggregation, and I'll outline the general case in the next chapter.

AT&T exemplifies another important lesson about disaggregation and revolutions. As AT&T disaggregated into smaller companies and the technology fell into separate hands, the bonds—interfaces—between the various pieces were carefully preserved. Local telephone companies still routed calls through the long-distance network; telephones that people purchased and installed in their own homes still were able to get dial tone from the local telephone companies. This didn't happen by accident—it was carefully planned. Successful disaggregations in technology, and the business of technology, repeat this pattern over and again: you must provide at least the basic pre-disaggregation functions of the technology (e.g., the ability to make phone calls), and to do this requires that you pay careful attention to interfaces between the pieces left behind after disaggregation (e.g., telephones still plug into the phone network).

Analysis

The goal of this book is to help you understand, create, and apply revolutionary technology. In the remainder of Part I, I'll discuss the details of disaggregation: how to understand what disaggregation does; what benefits to expect from disaggregation; and, given a problem, how to generate solutions based on disaggregation.

Chapter Two

Starting Revolutions: What to Take Apart

Disaggregation goes beyond simply taking things apart, and this chapter explains the details—how disaggregation works, what it does, and how to decide whether a particular innovation is important.

This chapter follows the same general outline I'd use to write up an analysis of an innovation. It's modified, of course—instead of discussing just one innovation, I've included some examples, and I've added all sorts of explanations. Still, if you want to analyze an innovation, write your own version of this chapter and you'll cover all the important points. (You'll need to read the next chapter, too.)

Here are the steps to analyze an innovation:

Step 1: Sort the Innovation

What is it, exactly, that the innovation disaggregates?

Step 2: Answer the Basic Questions

How does the innovation work?

Step 3: Assess the Revolutionary Potential

Based on the information from Steps 1 and 2, is this a revolutionary innovation?

In the next few sections, we'll discuss the details of how to analyze an innovation.

Step 1. Sort the Innovation

Someone hands you an innovation, and you think it works by disaggregation. Here's a question you need to answer: what is it—exactly—that's being disaggregated?

The *what* of disaggregation falls into five general categories. Innovations can disaggregate:

- Authority
- Ownership
- Mechanics
- Space/Time
- Concepts

Categories help suggest improvements. If we're trying to improve an innovation and make it more useful, we can ask questions like, "This innovation only disaggregates in category X. Can it disaggregate in category Y, too?"

Authority

Authority refers to the ability to determine what happens. If I manage something, if I control how it's made or what it does, if I decide how much it costs, if I control when something happens or if it happens at all, then I have authority. Authority can belong to a single person or to a group: a company, an industry, or a government, to name a few.

One important reason to disaggregate authority is to increase trust. A typical case is when manufacturers agree to follow some set of rules or an industry standard; they break off a piece of their authority and hand it over to third party, and in return they gain their customers' trust.

Another way to increase trust is to disaggregate authority and disperse it widely—to give authority to all participants. Sometimes disaggregation of authority is not about trust; authority becomes widely dispersed or simply evaporates, and the result is that people have more autonomy and freedom.

Table 2.1 shows an example of an innovation that disaggregates authority and increases trust.

Table 2.1: Authority

Before Disaggregation	**Medical treatment.** Individual doctors, acting on their personal judgment, make all decisions about end-of-life care.
The Innovation and What It Does	**Formal ethics process.** Hospitals assemble a team of experts to discuss ethical questions. Patients and their families participate in medical decisions.
Outcome	Patients and families trust that they will receive treatment in accordance with their wishes. Patient satisfaction improves when patients exercise autonomy in their health care decisions.

Ownership

Ownership has two meanings. The first meaning is the legal definition that refers to all the kinds of ownership that range from owning a shirt to owning stock market shares of a shirt factory. The other meaning of ownership is the one from the colloquial expression, where we say that some entity or some persons "own" something because they have a monopoly over it or possess it informally. For example, "Acme Corporation owns the market for computer memory chips."

An innovation that disaggregates ownership breaks the link between people and what they own. Usually the innovation will transfer ownership and it goes from one party to a different party. But there are times when disaggregation will make ownership just plain disappear—either whatever was owned disappears (for example, the market share vanishes), or it's still around but no one owns it. In more unusual cases, disaggregation forces a change in the *meaning* of ownership by changing the actual legal definition.

When disaggregation of ownership gives many people a stake in what's been shared, we often see a powerful result: more sharing. The more widely something is shared, the more people contribute. No one—or better yet, *everyone*—owns the World Wide Web, and millions of people spend time and money to put up informative Web pages just for the sheer joy of sharing.

Table 2.2 shows an example of disaggregation of ownership.

Table 2.2 Ownership

Before Disaggregation	
	One company owns the software. Traditional software companies own their product completely. The company writes the software, documents it, installs it, and provides technical support. The "source code," the instructions that let them build the software, is a carefully guarded trade secret. Customers do not own the software they buy; they only purchase the right to use the software.
The Innovation and What It Does	
	Open source software. Open source companies let everyone see the source code. Many open source projects go further and disaggregate ownership of the code—they give away the software for free and customers own their copies. Anyone can modify the software, and often projects allow people to resell the software.
Outcome	
	Everyone who uses the software becomes an owner. People share freely to improve the software: they contribute ideas, identify and fix problems, and write documentation.

Mechanics

Disaggregations in this category modify a mechanical relationship. What do I mean by the terms *mechanical* and *relationship*?

Let's say we're trying to build a better mousetrap. *Mechanical* refers to the physical properties of the mousetrap: the material that the mousetrap is made of, how it's put together, and any electrical characteristics. *Mechanical* also includes software, which is just as real as the nuts and bolts that hold the mousetrap together.

The word *relationship* refers to links, associations, and connections. Disaggregation doesn't need to split the mousetrap apart or do something *to the mousetrap itself.* Instead, the innovation can break the association between the mousetrap and some other object, like a specialized mousetrap-building machine. A disaggregation that breaks a "mechanical relationship" falls into the category of mechanical disaggregation.

Creative people like to tinker, and mechanical disaggregation often provides some very interesting "pebbles" to tinker with. After disaggregation, the pieces of technology are generally smaller, less complex, and more flexible than what was available before—and that's a winning combination when someone wants to be creative.

Table 2.3 shows an example of mechanical disaggregation.

Space/time

Innovations in this category take an event and break its connection to a particular place or time.

Innovations in this category are absolutely spectacular when they are aimed at people—at human events and the human senses. Over the past hundred years, disaggregation of some of the human senses—speech, sight, and hearing—completely transformed society.

Here's an ancient example: the art of writing. Writing lets people transmit their ideas to one another without the need to meet in person.

Disaggregations in this category produce two common results, which we'll see in many examples. The first is convenience: it's very handy to move events around in time and space, and many of today's most successful innovations provide that capability. The second common result is community: people use the ability to

Table 2.3: Mechanics

Before Disaggregation	**All the equipment on a factory floor runs off a single central engine.** In the days of steam engines, one big steam engine provided power for an entire factory. A mechanical transmission along the ceiling brought power to all the machine tools—lathes, drills, etc.—on the factory floor. Moving anything around meant fiddling with the transmission, which was difficult and expensive.
The Innovation and What It Does	**Electric motors for machine tools.** Henry Ford and his team were among the first people to use electric motors to power factory tools. They eliminated the rigid, hard-to-change connection between the central steam engine and the machine tools.
Outcome	Ford and his team gained terrific flexibility. They experimented with the machine tools, moving them around on the factory floor to find the most efficient layouts.

communicate with each other across distances and across generations to form communities.

Table 2.4 shows an example of the disaggregation of space/time.

Concepts

Sometimes an innovation takes apart a concept—a relationship that exists in people's minds. Concepts are more than just abstractions: our lives, our societies, and our institutions are built around certain ideas of how the world works. When a new innovation comes along and changes how we think, society also changes.

Conceptual disaggregation is at its best when it removes limits on how we think about problems. Creative innovations appear when preconceptions disappear.

Table 2.5 gives an example of conceptual disaggregation.

Table 2.4: Space/time

Before Disaggregation	**People held all conversations face to face.** The only way to converse with someone was to meet in person.
The Innovation and What It Does	**Telephones.** Telephones let people speak to each other no matter the distance between them.
Outcome	It's far more convenient to pick up the phone than to fly to New Zealand for a conversation. And the ability to talk to friends all over the world lets people form and maintain communities.

Table 2.5: Concepts

Before Disaggregation	**Money is based on scarce objects with intrinsic value.** People associated money with objects that had intrinsic value and were inherently scarce. Most Western civilizations used metals (gold, silver, copper) for money.
The Innovation and What It Does	**Modern currency.** Modern currencies separate the idea of value from the idea of intrinsic worth. Money is just an arbitrary token that people use to keep track of the value of other things, and it needs only to be scarce. United States dollars are valuable only because everyone agrees to use them.
Outcome	Economists realized that if money is just a token, and the government regulates the supply of tokens, then the government can increase or decrease the number of tokens to regulate the economy.

Step 2: Answer the Basic Questions

Now that we've defined all the categories, it's time to think about how disaggregation works. Here's the list of basic operational questions:

- Is disaggregation voluntary or involuntary?
- What is the scope of the disaggregation?
- Is the innovation technical, social, or a combination of both?
- What will happen to the pieces afterward?

If you're doing an analysis of an innovation, you'll want to ask all these questions over and again for each category. That's because the answers might be different in each category; for example, what happens to the pieces after disaggregation in one category isn't the same as what happens in a different category.

I'm going to discuss each of these questions.

Is Disaggregation Voluntary or Involuntary?

The question has a few different meanings. In the case of disaggregation of authority or of ownership, it's pretty straightforward: will the people who have authority or control surrender it because they want to, or because they have to? Sometimes people cheerfully surrender market share or authority; see Chapter 8. Other times they fight tooth and nail.

In the case of mechanics, the question usually is whether people will voluntarily adopt the innovation or not. After all, some people just hated the whole idea of the automobile.

In the case of space/time or concepts, again the same holds true. Sometimes people will cheerfully adopt the innovation. Other times they won't—I remember executives who demanded that all their e-mail be printed out and who never went online.

Just remember, not everyone will greet your new idea with a smile. This book is about avalanches, revolutions that change everything, and that means that some people will fight against change. It doesn't make a difference whether they lose anything—they just don't want the change.

What Is the Scope of the Disaggregation?

How far does the disaggregation go? Is it a full and complete disaggregation, or a partial disaggregation?

For example, when the government takes away authority, they rarely take it all away. The U.S. government has rules and regulations about safety, but companies have their own rules and regulations. Some of their authority is gone, but not all of it.

In other cases, the disaggregation is complete. When the government broke up AT&T, AT&T lost all ownership rights in the local telephone companies. Mechanical disaggregation—which includes taking apart relationships—also can be quite complete.

Is the Innovation Technical, Social, or a Combination of Both?

Although it's tempting to think that social innovations (business decisions, rules, regulations, laws, and court rulings) cause disaggregation of authority and that technical innovations cause mechanical disaggregation, it's not that straightforward.

Technical innovation can cause companies to lose ownership. A good example is that of the music companies; in the late 1990s, they effectively lost ownership of their product because of downloading via the Internet. Only later did the music companies really start to come up with sensible business models to adjust to the new technologies.

Social innovations can drive a mechanical disaggregation. For example, antitrust laws can force companies to redesign their products into smaller pieces, which allows competitors to provide some of those pieces.

What Will Happen to the Pieces Afterward?

If you take something apart, what's left behind? When AT&T came apart, all the pieces were left behind—after all, people still needed to make telephone calls.

Sometimes an innovation will take things apart but leave nothing behind. Ownership vanishes entirely when an artist puts a piece of art into the public domain. Mechanical disaggregation sometimes leads to more parts; but then again, a new design will often eliminate some parts—no single rule covers all possible innovations.

Step 3: Assess the Revolutionary Potential

Now it's time to step back a bit and look at the innovation as a whole. Maybe I'm convinced this is a good, solid, worthwhile innovation.

But is it revolutionary? Will it trigger an unstoppable avalanche? Is there any way to predict in advance? What makes an innovation revolutionary?

The analysis so far—sorting into categories and answering the basic questions—provides a way to let me predict whether an innovation is truly revolutionary. Well, more accurately, to *help* predict, because I still need to exercise some judgment.

Here are a few rules that help predict revolutionary potential:

- If an innovation disaggregates in multiple categories, it's more likely to touch off an avalanche.
- Innovations that produce multiple disaggregations—even in a single category—are more likely to start a revolution.
- An innovation that effectively provides for some basic human desire, or an innovation that performs a unique disaggregation in a particular category, is more likely to start a revolution.

We'll run across examples of all these situations in the rest of the book. As for what the "basic human desires" are, I include at a minimum the desire to share, the desire to form communities, and the desire for convenience—the desire to "make life easier."

The automatic teller machine, the ATM, is a nice, familiar, revolutionary innovation. Table 2.6 shows a quick analysis of the ATM.

The ATM does fit a couple of the rules. First, ATMs disaggregate in multiple categories; second, the ATM fulfills a basic human desire, convenience. Our analysis shows the ATM could be revolutionary, and so it was in practice.

I want to point out a bias in this book. Most of the examples I use are (I hope) pretty familiar to everyone. That means that whenever I bring up a revolution, I usually talk about one that everyone has heard of, a really big revolution that affects all of society. That's the bias—the idea that all revolutions are earth-shattering. What I want to point out is that a revolution can happen just *in a particular field* but still be revolutionary.

Not all revolutions are created equal. Sometimes an innovation will revolutionize an industry, but that doesn't mean that the innovation will change everyone's lives, or even that anyone outside the business will notice that it happened. Earth-shattering innovations don't

Table 2.6: Analysis of Automatic Teller Machines

Category	How the ATM Disaggregates
Authority	This category doesn't apply.
Ownership	ATMs are part of networks, which means that the banks **share** some ownership of their vital business function of giving me money and taking my deposits.
Mechanics	This category doesn't apply.
Space/Time	Money is available at places other than banks, and at any time of the day or night (at least in the United States). This provides incredible convenience that's extremely popular.
Concepts	I opened my bank account over the Internet; I get my money from ATMs; I deposit checks by mail. I've never even been to my bank's building, which I think is in Canada somewhere. I now realize that banks aren't big building with guards and vaults; they're a collection of business functions.

come along every day. A revolution might affect "only" an industry, or a single business, or a single department, or a single individual—but a small-scale revolution is still worthwhile.

Chapter 3

Benefits of Disaggregation: The Revolutionary's Bill of Rights

Chapter 2 explained how to analyze an innovation: how it disaggregates, what it disaggregates, how it works, and what basic human desires it fulfills. In this chapter, I'm going to discuss another important matter: what benefits to expect from an innovation.

The universal benefits of disaggregation are:

- Creativity
- Competition
- Cost Reduction
- Simplicity
- Specialization
- Synergy

I call them "universal" benefits because they show up, consistently, whenever an innovation works by disaggregation; every example and every case study in this book demonstrates at least some of these benefits. If you're a revolutionary—if you're working hard on innovations that disaggregate—then this list is your bill of rights. Your innovation should bring you at least some of these benefits, and possibly all of them.

Creativity

When disaggregation triggers an avalanche, it's not gravity that pulls on the rocks and gets them moving. It's creativity, pushing from behind, that sends the avalanche roaring down the side of the mountain. Creativity is the force behind the avalanche—creativity, as it finally escapes from behind the rocks that were holding it back.

What holds creativity back? Of course, there are lots of ways to prevent people from being creative, and probably a thousand books that explain how to free people's creativity, but I'll give my short version: creativity is stifled by authority, or is simply stopped in its tracks when people cannot figure out how to share. Disaggregation tears down these barriers and gives creativity a chance to escape.

And, of course, disaggregation increases the number of fascinating new and shiny pebbles for people to tinker with. If there's one thing that creative people enjoy, it's a new collection of pebbles.

Competition

After an avalanche smashes through the old forest, a new forest starts to grow. The new shape of the land lets plants grow where they couldn't grow before, the old trees that once choked out younger trees have fallen, and open spaces let seeds from far away take root and thrive.

Disaggregation sweeps aside the old infrastructure and changes the technological landscape. Companies that were once dominant can't keep newcomers away, innovation creates new niches and new ways to provide the technology, and people use the open space and breathing room to build new companies.

Cost Reduction

Disaggregation reduces costs—both the cost to manufacture and the cost to purchase. The way disaggregation reduces costs is usually straightforward. In the case of mechanical disaggregation, an innovation lowers manufacturing costs when a new design has fewer parts or when the parts become easier to manufacture.

When I analyze an innovation, I step back from time to time and take the broadest possible view of cost reduction. Voice mail is a good example of how an innovation saves more than money. If someone calls me and I have voice mail available, then the person saves time by leaving a message instead of calling back time and again to try and catch me. The caller also saves time because talking to voice mail is usually much faster than dictating a message to a secretary, and much more thorough, too. As for me, the receiver, I save time because I can hear all the details from the voice message, and I also save money because I don't have to hire a secretary. When voice mail first became popular, I used to hear complaints from people who didn't want to "talk to a machine," but today I hear complaints when there isn't a machine to talk to—because now people appreciate how the innovation reduces costs in their lives.

Simplicity

Once you've taken things apart, the pieces left behind—the pebbles—are simpler.

I'm not saying that this benefit, simplicity, necessarily makes *life* easier in the short term. When disaggregation changes one widget into ten new widgets, it's hard to juggle all the pieces and get them to work together. Fortunately, each new widget is generally simpler, and that's a measurable benefit: simpler to operate, simpler to improve, and simpler to repair. In technology, simplicity is a virtue, and disaggregation makes it possible.

Specialization

If you ever need to see a hundred or so examples of disaggregation, wander over to your local university library. Go the periodicals room, pick up a dozen or so trade journals from various industries, and start reading the advertisements. Each industry hatches dozens of highly specialized subindustries. Here are a few examples: a company that makes custom labels for wires to go inside electronic equipment; a service that screens potential call-center employees for their customer skills; a company in the publishing industry that checks computer files to make certain they'll print correctly.

The reason for all this specialization is quite rational: once disaggregation creates new pieces of technology, it's only natural that people decide to become experts in how the new pieces work. The converse is also true: people decide to become experts on some small corner of the technology, hang out a shingle, and before long another piece of the technology is disaggregated as experts compete against each other to reduce costs and improve service.

Synergy

After disaggregation, I've got a nice collection of pebbles. After an additional disaggregation, one that is perhaps in an entirely different field of business or technology, I've got yet another nice collection of pebbles. Synergy happens when I use pebbles from these different collections to create new products.

Here's a high-tech example of mixing pebbles from two different fields. The movie industry invented digital compression for music, and the computer industry invented wireless technology to let computers connect to networks without cables (another disaggregation). Put the two together, stir in a few audio components, and out comes a new product for home stereo systems: a little box that connects to the stereo through the usual cables and to a home computer through the wireless network. Whatever can be played on a home computer—and some people have quite a collection of online music—will play through the living room stereo.

Let's review the basic process of analysis, as presented in this chapter and the previous one. To evaluate an innovation, in addition to the analysis explained in Chapter 2, I run through the list of universal benefits and decide which of the benefits the innovation is likely to generate. For whichever benefits are likely, I provide a detailed description of the benefit(s).

A complete analysis of an innovation includes

- Sorting the innovation into proper categories
- Answering the basic operational questions about how the innovation will work
- Assessing the revolutionary potential of the innovation, including its likely scope
- Deciding what benefits are likely from this innovation

In the next chapter, we'll look at the opposite process: if you have a problem, how can you come up with an innovation to solve it?

Chapter Four

Four Stages to Revolution: Devise, Interface, Accept, Evaluate

Revolutions don't just happen; they have to be planned and executed. In this chapter, I'm going to discuss strategies and tactics to help accomplish the four stages of disaggregation, which are:

1. *Devise* your innovation.
2. *Interface* the disaggregated pieces so that they continue to work.
3. Get interested parties to *accept* your innovation.
4. *Evaluate* the outcome of your work.

Devise

In this first stage, you analyze your problem, generate a solution, and then carry it out. The trick is to construct those key ideas that lead to a solution, and here are a few methods to help that process along—ways to think about disaggregation that lead to insights into the problem and its solution.

Simple Inspection

Sometimes it's very easy to see how to achieve disaggregation. If my problem has to do with my company's structure, then my company is probably already divided into neat departments, sections, and divisions, and they even have convenient names like Shipping Department and Widget Solutions Division. If my problem has to do with computer hardware, I will find conveniently discrete items when I open up the computer case: disk drives, memory cards, and cables. I have no trouble imagining how to take a computer apart into separate components.

Nevertheless, don't stop with the first choice; if you stick with the obvious solution, you might miss deeper solutions or fundamental ideas. It may be tempting to split off the shipping department or the payroll function, but why stop there? Some companies have gone much further—they've outsourced pieces that were once considered core to the company. For example, in the semiconductor business, many companies that design and sell computer chips have gone "fabless"—they don't own their own "fabs," the highly expensive facilities that actually fabricate the chips. Such companies have decided that their expertise lies in design of the chips and in marketing, and they leave the actual manufacturing to the experts at the fab.

Restate Your Question

Try to restate your question in terms of disaggregation. There's a couple of ways to do this.

- State the problem in terms of breaking things apart.
- State the problems in terms of the universal benefits.

Let's say your current problem is, "My widget relies on gizmos from Acme, but deliveries from Acme are too unreliable." Here are a few ways to restate this question:

- "How do I end my reliance on Acme's gizmos for my widgets?" The focus is on how to break the link between your widget and Acme's gizmos. (Mechanical disaggregation.)

- "How can I get Acme to restructure their internal information systems to give me external access to their shipping schedules?" In other words, you want Acme to separate their proprietary information from the information you want them to share with you. (Disaggregation of ownership.)
- "How can I get Acme to sell me their gizmo patents?" Perhaps you can persuade Acme to separate the business of making money on a product from the business of manufacturing that product. (Conceptual disaggregation.)
- "If I were to have Acme build a larger piece of my system, would that reduce my overall costs?" Disaggregate your widget into various pieces, integrate some of the pieces with the Acme gizmo, and then perhaps the overall manufacturing and delivery costs will be lower. (Mechanical disaggregation.)
- "We need an off-the-shelf version of Acme's gizmos" reflects the quest for the universal benefit of commoditization. "Acme's lock on the market is making them careless" indicates the need for the universal benefit of competition.
- "Acme's gizmos are too general purpose" hints that you need the universal benefit of specialization. "Acme's gizmos aren't flexible enough" suggests that you need the universal benefit of creativity.

As shown in these examples, it's not just your own systems that might need disaggregation. It may be that your partners or suppliers need to disaggregate their technologies in order to give you access to their hardware, software, or processes.

If you restate the problem a few times, if you refine the problem down to its basic elements, you may find that you've stated the problem in terms of disaggregation. Listen carefully and see whether an implicit disaggregation is lurking in your basic problem statement. "Customers don't like our Web site" is too general, but it provides a useful starting point. Refine it again a few times and you might hear "Customers can see the product descriptions immediately, but the images take far too long to load." The key terms in that problem statement are *images* and *product description,* a split between the two types of data—text and graphics. When I hear the problem that way, I wonder whether it's time to reconfigure the Web site's computer system to handle text and graphics separately.

Consider the Categories

Another way to find a solution to a problem—whether it's a specific problem or a general quest for improvement and invention—is to frame a question and then look for a solution based one of the five categories of disaggregation: authority, ownership, mechanics, space/time, or concepts. Starting with the same question as before, "My widget relies on gizmos from Acme, but deliveries from Acme are too unreliable," we can reformulate our Acme problem. Here are some possibilities:

- "Acme doesn't inform us of its shipping schedules in a timely fashion." Perhaps it is time to propose joint ownership of the data. Joint ownership is a good way to encourage sharing.
- "Maybe Acme doesn't think we'll pay on time?" An escrow agreement can solve that problem by disaggregating authority over payment to a trusted third party. Disaggregation of authority leads to trust.

And so on and so forth in various combinations.

A simple example of stating the problem in terms of categories can be seen in the case of AT&T back in the 1980s. The U.S. government, among others, believed that the problem they faced was AT&T's almost complete ownership of the local and long-distance telephony market. When you formulate the problem that way, the solution is straightforward: the government forced AT&T to disaggregate into separate organizations.

Smash and Grab

Brainstorming provides an excellent source of new ideas, whether you're trying to solve a specific problem or whether you're simply trying to generate new and profitable ideas. I find brainstorming both very rewarding and very enjoyable.

To brainstorm a solution based on disaggregation, I recommend a technique I call "smash and grab": smash the existing links and grab the innovations that look valuable. Brainstorming in this way requires knowledge of the basic principles of disaggregation, a decent understanding of the field you're trying to improve, a flexible definition of what you want to achieve, and a willingness to play with each and every idea that comes along to see where it leads.

The technique consists of a walk-through. Imagine each step in the process, whether it's something simple like ordering a newspaper subscription or something complex like constructing a building. Then ask yourself along the way about the linkages, the connections, and the assumptions that seem to dictate that *this* step has to be connected to *that* step, or that *this* item must have a relationship with *that* item. Break off small pieces, large pieces, or entire functions and see where that leads you. Look to your competitors and your customers; break apart their technologies as well. Then try to recombine all the scattered pieces in different ways. How about if *this* piece were grouped with *that* piece? In the next section, we'll discuss the importance of connecting the disaggregated pieces together again.

One way to help move a brainstorming session along is to consider the categories along the same lines we mentioned just before. Brainstorm various ways to disaggregate authority, ownership, space/time, and the rest. Look at the universal benefits and apply them to your problem, even if they seem totally inappropriate. Remember, it's a brainstorming session, and no idea is rejected out of hand.

A colleague once asked me, as a challenge, to come up with an idea of how to disaggregate a restaurant. I had a few suggestions immediately, but a while later I decided to brainstorm the answer. I went through each of the steps in the processes of going out to a restaurant, from making the reservations to paying the bill at the end of the meal. I discovered that each step can be disaggregated to a greater or lesser extent. Given that restaurants have been around since the dawn of recorded history, even ideas that seemed somewhat outlandish to me (and that takes some doing) have existed someplace in practice. Can authority over what's on the menu be disaggregated? Certainly: the menu at a franchise restaurant is set by corporate headquarters. Can the ownership of the tables and chairs in a restaurant be disaggregated? Yes, and many malls in the United States actually have a similar arrangement: a "food court" with tables and chairs in a common area instead of separate sections for each restaurant. What about the meal itself—is it part and parcel of one restaurant's service, or can it come from multiple sources? A friend of mine ate a disaggregated lunch at a bazaar in India—he placed his order with a waiter who then gathered courses from several different restaurants to create a complete meal.

Assess and Strengthen Your Innovation

Now let's change the topic. So far we've discussed how to generate solutions—through simple inspection, through restating the problem, through thinking in categories, and through smash-and-grab brainstorming sessions. Before moving to the next stage in disaggregation, I strongly suggest that you analyze the disaggregation thoroughly. What categories does it fall under, and what will the effects be? Can it be strengthened or made more powerful by modifying it to fall under more than one category? Can you add additional disaggregations? Which of the universal benefits do you expect to see? Does your innovation fulfill any basic human desires?

Working in Parallel

As I stated in the beginning of the chapter, there are four stages to building an innovation: devise, interface, accept, and evaluate. We're almost finished discussing the "devise" stage, but before I discuss the remaining stages, I have a few comments on how these projects work in practice.

Work on the next two stages should happen in parallel. Actually, the first three stages—devise, interface, and accept—overlap to some extent. Here's a sentence that captures what I mean: "You must accept feedback in order to *devise* a solution with *interfaces* that are *acceptable* to the marketplace." Yes, you've devised your solution; but as you build interfaces with your industry peers, or solicit feedback from stakeholders, you'll find that you have to change what your innovation looks like. I admit that it would be nice to close the books at the end of the "devise" stage, but in practice you'll need to modify your innovation based on insights from the interface and accept stages.

Interface

Good, solid interfaces are crucial to any successful disaggregation. Unless you build interfaces, it's impossible to be certain that all the disaggregated parts can continue to work together.

Is it important that the pieces continue to work together? Absolutely. Disaggregation is very different than random destruction; after all, we want the technology to keep working! The ultimate goal of disaggregation is to provide at least the same functions as before but with a different infrastructure—to start the avalanche of change, but to still continue playing with the pebbles. If the technology doesn't have decent interfaces and if the pieces can't be made to work together again, then there's little chance that the disaggregation will succeed.

When AT&T broke up, everything continued to hum along quite smoothly—no lost telephone calls. The technical staff made all their interfaces work: they could connect the wires together and route the calls between the newly separate networks. The business relationships—in other words, the business *interfaces*—are just as important if you want to have a wire from Company A plugged into a socket at Company B. To define the business interfaces, AT&T business managers worked with the soon-to-be-independent companies far in advance. The net result of the efforts by the business and technical staff members to build both sets of interfaces was that the telephone calls went through—the original function of the telephone network did not suffer because of disaggregation. At the same time, the new and more flexible infrastructure triggered an avalanche of change in the telecommunications industry.

When companies neglect business or technical interfaces after disaggregation, people tend to notice. No doubt you've run into the problem of calling a company that's disaggregated, or that's built out of disaggregated parts, but is still working on the problem of building decent interfaces. Here's an example of a typical conversation with a company that doesn't have its interfaces fully in place just yet:

> "I need another widget, like the one I ordered last week."
> "Widgets come from a different department now, sir."
> "Can't you take my order?"
> "I'm sorry, sir. Here's the number, please call back."

Frankly, when I run across a company like that, my instinct is to find a different company with proper customer interfaces.

How much work is it to put interfaces in place? That depends on the innovation and who is using it; not all interfaces are the same.

Interfaces can range from informal, private interfaces that are visible only within your company to formal, rigidly defined and tested interfaces that are ratified by international standards bodies.

On the technical side, almost any company that builds electronic equipment will find itself defining interfaces, both of software and hardware, between the various modules that make up the equipment. These interfaces are usually private ones because there's no need for people outside the company to know anything about them. Two people building computer equipment in someone's garage don't need to go overboard defining architectures and reviewing documents—they don't need elaborate interfaces that adhere to the best practices of Fortune 500 companies. As a general rule, as the company gets larger, the documentation gets more elaborate.

On the business side, life is about the same. Small companies have almost no formally defined interfaces. Once past a certain size, business need formal interfaces—some dictated by business needs, others prompted by government regulations. Again, the larger the company, the more complex and extensive the rules.

Who makes their interfaces public? Companies that want to encourage the widest possible participation. A great example is eBay, which offers public interfaces as an essential part of their business strategy. Their goal was to create a thriving ecosystem around their company—and they've succeeded.

How did eBay do it? eBay disaggregated its internal computer systems—eBay split the data and systems that were eBay-only from the data and systems it wanted to share. Then eBay opened up its computers to outside companies by publishing interfaces—these interfaces let authorized computers outside the company interact with the computers at eBay. Companies develop software, register it with eBay, have it approved, and then sell it to their customers—the sellers on eBay. These professional sellers use the software to integrate their business with eBay's auctions.

Here's an example of how software packages can help sellers. Many companies sell on eBay to dump stuff that's been in inventory too long. These companies can use software to automate the process of putting old stuff up for sale: an application checks the company's inventory on a regular basis, and anything that's been in inventory too long goes to eBay's marketplace—automatically.

The result of eBay's efforts is a long list of clever software applications that enhance the value of eBay to professional sellers and greatly increase business for eBay. Recently some 40% of eBay's listings were generated through these software packages—and eBay didn't have to write the software.

Accept

Good interfaces are only part of the story. What must you do to get your innovation accepted? Just as before in our discussion on interfaces, the size of your innovation and the way it's used will determine your audience and what sort of campaign you'll need to get your innovation accepted. Theoretically, if your innovation is completely confined to your company, it ought to be very easy—right? All you have to do is get your managers, co-workers, factory-floor personnel, sales force, technical team, shareholders, and everyone else from the receptionist to the mailroom clerk to understand and accept your change. I'm not going to discuss how to introduce new ideas within your company; I refer you to one of the thousands of books on that particular subject.

When your disaggregation and the interfaces it creates are between your company and a limited number of other companies, a regular sales campaign should do the trick. FedEx, when it offers its outsourcing services to companies, sells its services as a way to save money and remedy problems. Wal-Mart, which wants its vendors to adopt specific interfaces, offers its vendors the opportunity for increased orders—or the prospect of no business at all if they refuse.

When your business model calls for thousands or tens of thousands of participants but it's impossible to spend time building a relationship with each and every one, you'll need a different strategy to gain acceptance. That was the challenge faced by eBay: sellers who use software packages generate the most revenue for eBay, so it's to eBay's advantage to make certain there are lots of good software applications for the sellers to use. How did eBay persuade the software community to accept eBay's interfaces and write software to work with eBay auctions? How did they develop the ecosystem of eBay, software developers, and sellers?

A crucial part of eBay's strategy was to use standards in its public interfaces. Using standards is like building a car: if you use standard nuts and bolts, then mechanics can use familiar tools to take the car apart and can then deal with the specifics of a particular model. eBay constructed its interfaces out of the standard "nuts and bolts" that software developers understand. Because eBay uses these standard pieces, software developers can use standard tools and techniques—tools that they understand and use anyway—to connect to eBay's interfaces. Developers don't have to waste time studying an entirely unique, eBay-only technology.

This isn't to say that eBay's interfaces are a "standard." They're "public," in the sense that companies outside eBay can use them and anyone who is interested can read them—but they are not *standard,* any more than a car built using standard nuts and bolts is a "standard" car. eBay created the interface, eBay owns it, and the interface hasn't been approved or endorsed by an industry group or by a national or transnational authority. To date, eBay hasn't felt the need to adhere to any of the proposed international standards for business interfaces.

Standards may help you get your innovations accepted by the marketplace. Standards can help you in one of two ways: you can incorporate standards into your innovation as eBay did, or you can make the innovation itself into a standard. Or, for that matter, you can combine the two approaches by incorporating previous standards into a new standard. That's part of the avalanche effect: a disaggregation, which is what a standard fundamentally is, accelerates change and revolution by enabling other standards.

If your innovation affects everyone, if it will be as ubiquitous as the Internet, then you have to sell it to everyone and it is likely that standards will help. A standard disaggregates authority—to transform your interface into a real standard, you must cede at least some control over your innovation to a standards body. What you get in return is trust: customers believe that your innovation will reflect the interests of the entire community, not just the interests of one company. Chapter 8 discusses the role of standards at length.

Even if only a handful of companies are involved in your innovation, sometimes standards are still the way to go. Especially when you deal with your competitors, you need to build trust; and when you need to build trust, it may be time to disaggregate authority. To introduce music CDs, the electronics industry and music industry had to cooperate with each other, one to produce CD players and the other make CDs to play. The CD revolution was made possible by a standards body. Standards bodies offer a generally democratic process to resolve differences and smooth the way to mutual cooperation.

By way of horrible example, the DVD industry is (at the time of this writing) in the middle of a vicious war between different—and *incompatible*—DVD formats for the next generation of video DVDs. The industry has successfully persuaded me to avoid the next generation of DVDs until it has all been settled; I somehow doubt that's what they intended but it is the inevitable result. If you want me to put money down for a new player, you have to make me believe that the player I purchase won't become useless in a couple of years—that I can trust the industry to support the player technology I select. And I certainly won't buy DVDs in the new format either until I know that I'll still be able to play them in ten years.

In a perfect world, my reluctance to buy a new DVD player should provide feedback to the industry—they disaggregated authority but did so in a way that didn't build trust. Unfortunately, the media industry has a long and glorious history of shooting itself in the foot instead—I'll discuss this fiasco in Chapter 8.

Building Your Innovation

At some point in this cycle of devise, interface, and accept, the time comes to "actualize" your innovation. I won't discuss how to make your innovation work in practice—an innovation can be anything from a corporate divestiture to a new kind of wristwatch. But I would like to offer one observation: the act of building an innovation provides important feedback to the stages of devise, interface, and accept. Prototypes, trials, field tests, war games—however it is that you test your innovation—provide data, examples, and ideas that can support you as you move through these three stages. Although I haven't mentioned implementing the innovation until now, that doesn't mean you should leave it for last; in fact, the earlier you get started, the better. You will find that building your innovation is a valuable part of the feedback cycle.

Evaluate

You've devised the disaggregation, you've built the interfaces, and they've been accepted in the marketplace. Your innovation, technical or social, or both, is real. Clearly it's time to kick back, relax, and enjoy a job well done. That's fine, but afterward it's time to get back to work again and evaluate how well your innovation works, what it's doing, and what you should do next.

Disaggregation, if properly handled and sufficiently powerful, can trigger an avalanche. Disaggregation can reshape the landscape and sweep everything out of its path. It's crazy to stand by and say "Gosh, that was interesting" and then walk away just when the avalanche is starting, especially if your innovation started it in the first place. When your product introduces a powerful disaggregation, it's time to understand the implications and evaluate what to do next. Did you see the benefits you expected—the ones you anticipated when you devised the innovation in the first place? What were the unexpected side effects—benefits you didn't expect and drawbacks you didn't anticipate? How are you going to fix remaining problems? What are your competitors going to do? What pebbles of technology were unexpectedly revealed by the avalanche, and who's going to pick them up and use them?

Twenty years ago we'd draw a project as a nice, neat linear process. Flow charts connected stages of the project in an orderly fashion, and the flow charts I saw always ended the project with delivery to the customer. These flow charts had a little problem: they were utterly bogus. No project was ever that neat.

Industry abandoned this "start, middle, finish" fantasy; today, we treat projects as cycles. The development process is a wheel that keeps on turning. When you're "finished," that is, when you've got the innovation out the door, it's time to begin again and create the next revision to incorporate improvements, lessons, and feedback from the first time through. The Internet isn't interesting, powerful, and exciting just because it was a fabulous idea—the Internet is all of these things because it's a continuous revolution, with people working nonstop to add improvements.

Should you celebrate the day your product ships or your industry standard is accepted? Absolutely. You've created something new and special. You've done a great job and deserve a reward. Just remember: no matter how good a job you've done, no matter how profitable or how excellent your finished product, technology never stands still, and neither should you.

An Important Distinction

Before I sum up, I would like to point out an important distinction. In this chapter I've discussed the idea that problems can be solved with disaggregation. From time to time, *the fundamental goal you're trying to achieve* will turn out to be disaggregation. In that case, the solution to your problem can be very tricky because the solution may require an entirely new technology.

Thomas Edison faced this exact problem. Edison wanted to invent a way to let people listen to speech and music without being present when the words are spoken or the music is played; in other words, he wanted to disaggregate in space and time. He and his team invented the phonograph, but the phonograph wasn't the *result* of disaggregation—Edison didn't take existing technology apart to build the first phonograph. Instead, the phonograph was the product of deliberate, painstaking creation of new technology—Edison *achieved* disaggregation with the phonograph.

But in general, it's important to remember that disaggregation can generate further disaggregation—the avalanche effect. Take the example of Gutenberg. Before Gutenberg, every page in a book required its own custom-made printing plate, which was expensive and time-consuming to fabricate. Gutenberg wanted to break the association between the printing plate and the page it printed. He came up with movable type, which disaggregated each individual letter from the plate. With movable type, each letter could be added separately to make up a printing plate, which resulted in dramatically lower costs and rapid turnaround times. The mechanical disaggregation of letters from plate created the more important disaggregation of plate from page.

In this chapter we discuss the four stages of disaggregation. First, we discuss how to *devise* an innovation, including methods for generating ideas. This stage overlaps with the next two stages, *interface* and *accept,* which we work on simultaneously. In the *interface* stage, we build interfaces for the disaggregated parts, which must continue to work together; of course, these interfaces might also provide entirely new capabilities. In the *accept* stage, we persuade our audience to accept the innovation and its interfaces. This leads to a feedback loop between the stages of devising the solution, creating the interfaces, and gaining marketplace acceptance: "You must accept feedback in order to *devise* a solution with *interfaces* that are *acceptable* to the marketplace." When all is said and done and the innovation is out the door, it's important to *evaluate* what's been built—to keep an eye out for the oncoming avalanche. You wouldn't want to be swept away by an avalanche of your own creation!

Case Studies: Two Hundred Years of Revolutions

Part II

Part II discusses three revolutions that happened between the late 1700s and yesterday morning.

In the 1800s, the technology of manufacturing completely changed as handmade crafts gave way to the assembly line.

In the 1900s, the automobile replaced the train.

In the 2000s, Internet technology introduced a powerful new tool that transformed business and technology.

Each of these revolutions introduced startling changes in our everyday lives—and, for that matter, none of these revolutions are over just yet.

Chapter Five

From Horses and Buggies to Jet Planes: The Revolution in Manufacturing

Where do all these machines come from? My house is full of machines, from little clever-but-simple ones like retractable ballpoint pens, to medium-size ones like paper shredders and microwave ovens, all the way to full-size ones like recliners and washing machines. My house is partly a machine: the walls are full of machinery (pipes, switches, cables, and ducts), and so is the roof, with its vents and fans. The furnace, water heater, and air conditioner are hidden in the basement.

I've visited reconstructed homes of the late 1700s, and there's nothing remotely comparable to what we have today: the most intricate machine in the house is a technology that dates from the Middle Ages—a spinning wheel sitting in the corner. Furniture drawers open and close, and chest lids can be raised and lowered. The door has a metal hinge; that's the most complicated bit of machinery built into the house itself. These houses were bare of almost anything we think of as a machine. The people of the time certainly could and did build complicated machines, but none of them were in ordinary people's homes. How did our modern homes fill up with today's wonderful machines?

The story starts with what is arguably the most important innovation in manufacturing history. In 1798, Eli Whitney introduced a brilliant new manufacturing technique to make guns for the U.S. Army. Before Whitney, each gun, like all manufactured goods of that era, was a unique product; craftsmen painstakingly created each gun individually. The parts of any particular gun—the barrels, the triggers—had unique dimensions and fit together only in that particular gun. Whitney built guns with interchangeable parts. Each trigger was identical to every other trigger, each barrel was identical to every other barrel, and so forth. To build a gun, you basically started with a pile of parts and fitted them together.

Whitney started a revolution with this mechanical disaggregation of guns from their component parts—and soon everyone wanted to adopt his method of manufacturing. What made Whitney's innovations so compelling—what were the benefits to Whitney, and what were the benefits to his customers?

One reason for Whitney's success was that customers gained enormous benefits from interchangeable parts. A constant headache for armies is keeping guns in repair. If each gun is an individually crafted masterpiece, repairing the gun requires a highly trained specialist. By contrast, modern guns have interchangeable parts, and you can grab a replacement trigger out of a parts bin. If a soldier has a couple of broken guns in front of him, he can cannibalize parts from one gun to repair another and get a working firearm.

Another benefit was the central supply of bullets. When every gun was different, every gun needed bullets sized just right for that gun, which was a logistical nightmare in wartime. Militia soldiers called to the colors had to bring their own bullet molds with them so that they could make bullets for their guns. But once all the guns in the army were assembled from identical parts, they all fired identical bullets, and bullets came from a central depot.

Manufacturers became interested in Whitney's ideas because Whitney made money. Whitney reaped an important but typical benefit of disaggregation, cost reduction. This was no accident—Whitney realized that disaggregation would lower his manufacturing costs. This left the important question of how to actually make identical parts. After all, when pre-Whitney gunsmiths made guns, they didn't deliberately try to make them all different. They used the same basic measurements to make each gun, but their *processes* and their *tools* did not produce interchangeable parts.

Whitney is the inventor of the modern tool-and-die industry. Tools and dies are special machines that are each designed to help you build particular parts. Place a piece of metal into the die *here* and cut it off so it aligns *with that mark*; bend it until it fits into *that curve* exactly; now you have a piece of metal exactly the right length, bent to the exact shape you need. Not only did tools and dies allow Whitney to manufacture parts to the correct tolerances, but they also meant he could use ordinary skilled workers in his factories instead of gunsmiths (who required years of training and who commanded higher wages). As a result, Whitney's costs dropped at the same time that his customers received a better product.

Because sloppiness doesn't go well with high explosives, guns require precision parts and careful alignment. Whitney's processes produced identical parts but his workers still had to fit the guns together very carefully—in other words, guns didn't suddenly become cheap. Whitney built an expensive product for a client who was relatively insensitive to costs—no matter what, the army had to have guns, and guns were expensive. What about a something cheaper? Something for the mass market? Maybe Whitney's ideas weren't suitable for ordinary, everyday stuff?

Of all things, it seems that wooden clocks drove the avalanche onward. A clock maker by the name of Eli Terry adapted Whitney's methods to the production of wooden clocks. Not only that, but Terry took another big step: he introduced a new conceptual disaggregation into clock manufacturing, and, as often happens, the additional disaggregation increased the power of the avalanche. In 1807, Terry signed a contract with a merchant to deliver four thousand clocks in three years' time—the first instance in which the business of *selling* clocks was disaggregated from the business of *producing* clocks. Four thousand clocks was an absolutely staggering number, more than a traditional clockmaker would make in a lifetime, much less three years! Terry adapted Whitney's methods to his particular problem.

When I say "wooden clocks," I really mean it, by the way: wooden almost all the way through. The gears were made of wood, the case—if there was one—was made of wood, and, because the first generation of clocks operated by a system of weights, they didn't require an expensive spring. As long as the gears meshed properly, they didn't need anywhere near the precision required for guns. Terry designed clock gears with teeth that meshed even if measurements were slightly off. Assembling the clock included some adjustments to make certain that it kept time correctly. The result was a clock with interchangeable parts—inexpensive interchangeable parts that did a perfectly adequate job of timekeeping. Terry was successful—wildly successful—and other clock manufacturers had to adopt his methods simply to survive. The processes of mass production—tools and dies, and interchangeable parts—spread throughout the United States's wooden clock industry. An avalanche of change came roaring through the industry and swept aside the old methods of manufacturing.

Clock manufacturers gradually improved their processes, and the individual clock parts became more precise. Better precision let them introduce another disaggregation: if clocks are assembled from precision interchangeable parts, they don't need to be manufactured by a single company. As we learned in Part I, you can't just take technology apart; you need interfaces to let the disaggregated pieces of technology continue to work together. In the case of wooden clocks, the interface was an industry standard for the dimensions of the gears. The manufacturers invented clever and inexpensive ways to measure their products to make certain they met the standards. The industry began to divide up into specialties; some manufacturers would build the parts of the clock, such as the gears, and other manufacturers bought those parts and assembled them into clocks. Even more interesting, parts from different manufacturers fit together. Now different companies owned different parts of the process. The business of producing the parts disaggregated from the business of assembling the parts into clocks.

Now's a good time to pause and ask what some of the lessons are. The primary lesson is the same one I present throughout this book:

disaggregations start avalanches—revolutions that sweep aside old technology. If you combine multiple categories of disaggregations, the innovations are even more compelling and the avalanche has even more power. Mass production of wooden clocks did include multiple categories of disaggregation; here is a list:

- The basic building block of mass production was the mechanical disaggregation of parts from the final product.
- Disaggregation of authority let people agree on interfaces between the various clock parts.
- People created separate businesses to manufacture clock parts, and no one company owned the entire business any longer; in other words, they disaggregated ownership.
- Terry introduced conceptual disaggregation into the clock industry when he separated the business of selling clocks from the business of producing clocks, and at the same time eliminated the notion that clocks had to be produced by craftsmen.

There's another lesson in the story of mass production, and in fact it is the reason I chose this example over others: sometimes it's very easy to start the avalanche. No one had to force manufacturers to switch to the new Whitney/Terry methods. Every manufacturer could independently decide how their factories would work, and not surprisingly they all decided to adopt the new lower-cost methods—or, at least, the survivors did. The ones that didn't went out of business. So the changeover to new methods of manufacturing was a natural process of individual decisions made for sound economic reasons.

No discussion of disaggregation is complete without thinking about the how the innovation provided at least some of the usual set of benefits. Let's discuss cost reduction, competition, specialization, and creativity.

We already discussed the how manufacturers trimmed their costs. Society as a whole benefited as the cost to consumers plummeted. When Terry started his clock business clocks cost about $40 retail—there were no clock wholesalers because the concept did not exist—and in ten years the retail price dropped to about $16 with a wholesale price of about $12. By mid-century, when wooden clocks were finally replaced by metal clocks, they cost only thirty-seven and a half cents wholesale. Clocks went from a rare luxury item to a common household good that everyone could afford. (I'm not going to debate whether having a clock is a good idea; clearly, people wanted clocks and found them useful.)

With specialized machinery available, Terry's new method of selling clocks firmly established, and a huge untapped market for low-cost clocks, competition became quite fierce—but in a curiously collegial way. Since anyone could dissect these simple clocks to see how they were put together, the manufacturers were very open with each other about their designs and methods. The foremen of various clock companies would visit each other to share ideas and innovations, and the best ideas spread from manufacturer to manufacturer.

Disaggregation encouraged creativity in several ways. First of all, of course, everyone had to be creative and attempt to produce a lower-cost or more attractive-looking clock. Even more fundamentally, the disaggregation of the components of the clock let the manufacturers think about how to produce a single component more efficiently. If you're building clocks one at a time you might think about how to build a clock more efficiently. But if you're building thousands of identical gears, you'll start to think about improving gears—a small cost reduction will be multiplied a thousand times over. Records survive from one clockmaker who kept careful track of how long it took to produce a thousand crown gears (seven days) or drill out a thousand lift shafts (just one day). How about a better way to saw the gears? What about making the clock slightly less expensive to assemble? People started asking, and answering, these questions.

Finally, there was the benefit of specialization. With the parts disaggregated from the final clock, some manufacturers began to specialize in making parts while others specialized in assembling them. This meant that each could focus on doing its particular job more efficiently. Although I haven't run across evidence for them just yet, I'll bet that little niche businesses sprang up around helping the component makers and assemblers do their jobs more efficiently, just as they do today. This kind of creativity is part of the essential nature of disaggregation; once someone gets the idea that disaggregation is useful, they continue to disaggregate and introduce new and useful innovations.

How did our modern homes fill up with today's wonderful machines? The falling prices of wooden clocks offered a lesson that manufacturers of other goods could not ignore—in fact, it seems that they were happy to apply those lessons to a wide range of products. More and more complicated machines began to appear in people's homes; eventually, even highly complex and very expensive machines, such as the sewing machine, came within the reach of ordinary people. The inventions of the late 1800s, which included such marvels as the phonograph, could spread more rapidly because mass production made them relatively inexpensive. The home of the year 1900 was a place very different from one of the late 1700s, and we can give a great deal of the credit to Eli Whitney—and to disaggregation.

Chapter Six

The Automobile Takes On the Railroads

Some innovations focus on a single great idea, execute it well, and completely overturn an established industry. This case study is about one such innovation, the automobile.

The Industrial Age: Canals, Railways, and Automobiles

At the beginning of the Industrial Age, it was difficult if not impossible to transport the output of a factory across any significant distance; horse-drawn carts couldn't keep up with the tremendous output of a large factory. To help solve this problem, England and the United States built canals to move goods, but barges are relatively slow, and canals can't reach everywhere.

The invention of the railroads made mass production useful—the output of a factory could travel quickly and economically to anywhere there was a railroad, and railroads can go just about anywhere. Besides moving goods, railroads offered another terrific benefit, the ability to move people safely and reliably. Before railroads, a journey between two cities wasn't just a trip, it was an adventure—an adventure that not everyone survived. After railroads, a cross-country journey soon became as simple as buying a ticket and hopping aboard a train.

To this day, because of their importance in moving industrial goods, railroads are a crucial measure of national industrial power. Railroads also provided another form of national power: military power. In the 1800s, the military quickly learned how to use the railroads to move troops and their supplies over great distances. Between the industrial value and military value, railroads were paramount in the 1800s. The question of which European country had the right to build a railway in an Eastern or African colony nearly started a war or two.

In the late 1800s, the automobile introduced a simple mechanical disaggregation: machine-powered transportation, but without the rails. I say "simple" only because it was a simple idea—a lot of hard work went into making it practical. At first the automobile was very expensive, but in the early 1900s, Henry Ford's innovations reduced the cost of the automobile and put it into the hands of the general public. The benefits of the automobile, which we'll discuss below, became available to the masses; the general public got behind the innovation and pushed, and the avalanche was unstoppable. The innovation overturned the old industrial order, and the age of the railroad gave way to the age of the automobile.

The automobile disaggregated, layer by layer, the technological and social infrastructure that the railroads created to support their technology. I'm going to discuss each layer of the infrastructure of the railroad, how the automobile disaggregated that layer, and the benefits that disaggregation of that layer has provided.

The Roadway

Railroads need tracks, but automobiles don't. This is the most obvious disaggregation, a simple mechanical one, but one with terrific benefits nonetheless. Railroad tracks are difficult to design, hard to build, and expensive to maintain. The rail bed can't be too steep, can't curve too sharply, and can't tilt too much from side to side. The automobile breaks this strong association between the vehicle and its roadway. The consequence of mechanical disaggregation is flexibility, and automobiles can travel almost anywhere—on roads chopped out of ice or along tracks in the hot desert sand, down muddy streams, and up rocky hills. Using various types of materials—asphalt, concrete, gravel, dirt, steel—engineers construct everything from backcountry dirt roads to multilane superhighways.

There's more here than terrific cost savings. Because automobiles don't require an expensive roadway, automobiles are easier to integrate into a community. Maybe your small town can't support a railway track, but it certainly can support a dirt road. Many a quiet backwater that will never see a railroad has been visited by automobile. Furthermore, an automobile road is infrastructure that everyone can share. Railroad tracks are obstacles, really; if you have railroad tracks all you can do is run trains on them, and all other traffic requires crossings or bridges across the tracks. Automobile roads can support every kind of traffic—pedestrians, bicyclists, and ox-drawn carts.

Although it's somewhat tempting to stop our analysis with the purely mechanical disaggregation, I chose this case study because it has wider lessons for some of today's technology. When I assess an innovation I think of the implications for the *entire* infrastructure of the technology—and railroads, both then and now, are more complex than they appear at first glance. I'm going to discuss some of the less obvious implications of the automobile, and I'm going to start with two social results of disaggregation.

The first result is that because automobiles travel on public roads, almost anyone can take his or her own car and get on the road—a big change from the railroads with their severe limitations on access to the railways. To put this in terms of disaggregation, automobile roads are shared by everyone—disaggregation of ownership. The automobile's easy access to the roads means competition and cost reduction; no company can shut other companies out of the package-delivery business if all you need to compete is access to the roadway, a truck, and a driver.

The second result is that unlike the railroads, the decision of where to put an automobile road is usually a more open process—a disaggregation of authority. Railway tracks are a very scarce resource; who decides where the tracks go? A private corporation might make the decision; if the railroad is run by the government, the decision requires high-level political and economic maneuvering because of the high costs. Most automobile roads are local public property—communities decide for themselves how extensive, and how expensive, their automobile roads will be.

The Traffic

The traffic on the railroad—the railroad cars—hook up to each other and the locomotive to form trains. The railroad cars have mechanical couplers to attach to the next car, hydraulic connections for the brakes, and maybe electrical connections. The traffic on streets and highways—the automobiles—don't hook up to each other; one of the nice benefits of the automobile is that each one works independently. This independence would seem to imply that the automobile has no interfaces to other automobiles.

But that's not the entire story because the automobiles must still work with each other—automobiles must cooperate on the roadway; otherwise, they'll crash into each other. Automobiles, *because* they are disaggregated from each other, need interfaces to each other. Table 6.1 lists the interfaces used by railroads versus the interfaces used by automobiles.

I would like to point out one curious detail. Automobiles have an *aggregation* that railroad cars don't have: the engine that pulls the automobile is inside the automobile. Railroad cars don't have engines—the engines are in the locomotives; in fact, the disaggregation of motive power from the individual railroad cars is what gives railroads their enormous efficiency. Tractor-trailer trucks use the same model as railroads: a locomotive (the tractor) pulls a cargo vehicle (the trailer). Just like the railroads with their locomotives and railroad cars, truck tractors and trailers use standard couplers and connectors to hook together. Mechanical disaggregation of motive power from the vehicle means that there must be an interface, and that's true regardless of whether the vehicle travels on railways or on roads.

The automobile's disaggregation from the roadway and from other vehicles translates into the expected benefits of disaggrega-

Table 6.1 The Interfaces of Railroad and Automobile Traffic

Railroad	Automobile
Railroad cars hook up to each other with mechanical, electrical, and hydraulic couplers.	Automobiles signal each other with turn signals, horns, and headlights.
Rolling stock must be able to run along tracks, around curves, and through tunnels, which imposes strict rules on the size, shape, and length of the railroad cars.	Automobiles have much more lax constraints. The restrictions they have are mostly due to the height of bridges and the need to navigate narrow city streets.
Equipment for loading and unloading railroad cars must interoperate with the cars or else the cargo won't load.	Automobiles get fuel from gas pumps; gas tanks and nozzles must be compatible with each other.
Trains automatically trip signaling equipment, which sends signals to other trains.	Traffic lights control the flow of vehicles. Some roads have sensors in them to measure traffic, but the automobiles have no interfaces with the road—at least, not yet.

tion; the most obvious ones are creativity and specialization, which you can see by looking at the traffic on any busy street. Small, low-slung sports cars share the road with huge tractor-trailer trucks; family vans drive alongside gravel trucks. Businesses use oddly shaped vehicles for specialized tasks, both industrial (e.g., glass delivery) and advertising (e.g., the Oscar Mayer Weinermobile). There are motorcycles and motor homes. Vehicles that do not travel the public roads have even more latitude in their design because they don't have to obey size, height, width, and speed restrictions: children drive small motorized cars in their backyards, while mining companies use truly gargantuan trucks to carry ore. The single innovation, the disaggregation of mechanized transport from roadways, made all this possible.

Network Management

Railroads are *managed,* to a really amazing degree. In the 1990s, the Union Pacific railroad in the United States acquired the Southern Pacific railroad. Union Pacific promptly went on a typical corporate-America postacquisition firing binge and smugly ignored the advice of the remaining Southern Pacific personnel—the only people who actually understood the policy, procedures, customers, and traffic patterns of their railroad. The result was complete and utter chaos: thousands upon thousands of railroad cars were lost or misdirected. The railroad yards and even the railway sidings jammed up with idle trains. The situation got so bad that trains ran out of fuel as they were trying to maneuver along the tracks! It was an impossible mess; the traffic snarls had a major impact on industrial output all across the United States.

Here's a quick list of some of the management challenges you must solve when you run a railroad:

- Intercity rail lines often have just a single track but trains must travel in both directions. At a particular time and place, only one train can use that piece of track.
- Trains must keep careful intervals for safety's sake. A train traveling at fifty-five miles per hour has a stopping distance of over a mile.
- Locomotives are expensive, and the railroad cars can't move themselves. Trains are collections of locomotives and railroad cars, and the economies of scale dictate that trains be as long as possible.
- Railroad cars must be switched between one train and another to get the cargo to its destination.

Railroads solve these problems through meticulous central planning. The legendary railroad timetables were the first exercises of their kind in nationwide planning and required strict, by-the-minute timekeeping.

Automobiles require nothing of the sort—which is part of their charm. Automobiles are personal transportation. When I want to go someplace, I get into my car and go. I don't have to check with a dispatcher in downtown Chicago to clear my trip to the grocery store, and if I suddenly get the urge to drive to Montana I don't need anyone's permission. My car's path through the road network, its schedule—

and in fact whether it arrives at its destination at all—is my business alone. The owners of the road attempt to regulate the flow of *traffic* using streetlights, stop signs, and speed limits, but the responsibility for any kind of transportation *network*—the delivery of cargo to its destination—is in the hands of whoever wants the cargo delivered.

In other words, automobile disaggregates ownership of network management from ownership of the roadway. Disaggregation leads, in turn, to the usual benefits: creativity, competition, and specialization, all in the area of managing transportation networks. Transportation companies have created many different ways to build networks and manage the flow of cargo through the public roadway system; specialized logistics companies work on building better systems. Since better systems bring better profits, there's been a steady improvement in how transportation networks are managed.

An example of the benefit of simplicity offered by the automobile: automobiles move independently, not in "trains." Let's consider the problem of delivering one hundred loads of Pez candy dispensers to one hundred different stores. Railroads put the loads on trains, move the trains around as a unit, move the loads between trains as necessary, and eventually deliver the loads. If delivery is by truck, the trucks more or less just take off for their destinations, each one following its own path. (Some shipping companies do dictate paths for their trucks; it's a decision that each company makes for itself.) Each truck driver optimizes his or her own path through the roadway system and makes choices at each intersection. Right turn or left turn? Expressway, tollway, streets? Bridge or ferry? If all one hundred trucks leave one after the other for the same destination, the order of their arrival isn't guaranteed to be the same as the order of their departure. Although it sounds messy, it means that each driver can optimize, adapt, and decide what to do, based on local knowledge of current traffic and weather conditions. Distant managers don't have to dictate and coordinate each decision. It's a glorious confusion, and practical too.

The Business

A railroad company is a business, and, like any other business, each one has ideas about how to treat customers, what services to offer, how to turn a profit, as well as all other questions big and small. Let's say I'm a customer of a railroad and I want or need a new kind of service—faster delivery, special handling of my cargo, or maybe I don't want my cargo to move on the Sabbath. Unless I can convince the railroad's management to offer the service, it won't happen; and don't forget that the bigger the enterprise, the less attention management pays to "small" ideas because they don't contribute much to the overall bottom line.

If the railroad company won't accommodate me, is there any way to get the new kind of service I want? One benefit of disaggregation is competition, and since (as we've just learned) automobiles disaggregate the business of transportation from ownership of the road, I expect competition in the trucking industry and I can reasonably expect to take advantage of it. If one company turns down my request for a new kind of service, another one might be willing to listen. If I persevere, it's likely that I can find someone to give me the service I need.

Another benefit of disaggregation is creativity. If you come up with a wacky new idea that uses automobiles, you can simply start the business (assuming, of course, that you're not violating any existing traffic laws). Innovators have created all sorts of amazing "applications" for automobiles, some of which constitute entire new industries. Package delivery, mobile medical care, taxis, pizza delivery . . . ; the list is endless. And many of these industries have spawned sub-industries. Consider package delivery. Package delivery includes, in part, overnight delivery, express delivery, messenger services, freight delivery, heavy equipment transport, and live animal transport. The freedom to create, to compete, and to specialize spawns new business almost every day, all based on the automobile.

Like any good avalanche, the automobile triggered a series of other avalanches. Entire industries revolve around the automobile: repair services, insurance companies, the oil industry, and driving schools, to name a few. And then there's the larger question of how the automobile changed the patterns of our daily lives. For example, consider this disaggregation enabled by the automobile: the place where we work no longer needs to be near where we live—which makes possible the city/suburb model of living. All of these changes and questions are interesting, but they are for another day.

The fundamental innovation, taking motorized transport off the rails and putting it onto the roads, had a profound effect on the transportation industry. This simple mechanical change disaggregated the industry, layer by layer, and introduced disaggregations in ownership and authority. The changes went beyond just the machinery and included the very structure of the transportation industry. The innovation triggered an avalanche of change, delivered all the benefits we expect from disaggregation, and continues to provide a source of interesting innovations to this day.

Chapter Seven

The Internet's Permanent Revolution

The Internet operates in a state of almost total anarchy. My computer defenses record upwards of a hundred attacks a day against my office network by viruses, crackers, and other malicious intruders, and that's a drop in the bucket compared with some systems. Then there's spam. Over 80% of the e-mail I receive is spam, and if it weren't for my antispam filters I would have given up on e-mail long ago.

Security and privacy are your problems to solve when you go online. As for getting online in the first place, Internet service providers (ISPs) aren't common carriers and they don't have to carry your data if they don't want to, and, even among themselves, the ISPs trade data under a series of ad-hoc agreements. Anyone can send any kind of data they please, anyone can connect to the Internet, and governments find it almost impossible to censor or restrict Internet access. For a surprisingly long time, the *root name server* for the Internet—the supreme worldwide authority that translates alphabetic names like "www.example.com" into the series of numbers that computers actually use—was an ordinary computer underneath someone's office desk.

It wasn't supposed to be this way. The worldwide data network was supposed to be carefully managed and run like a Swiss watch, built to specifications formulated by an international telecommunications agency sanctioned by the United Nations. The large national telephone companies, which already owned all the wires that carried data, would carefully shepherd data over the network. There might be up to ten computer networks in each country to carry data (those wild-eyed capitalists in the United States thought they might, one day, have as many as one hundred different networks!). Data would travel according to predictable and stable patterns.

Instead of the official UN plan, we have a permanent revolution: every day brings fresh ideas and new ways to use this marvelous communication tool. Instead of a dozen networks, there are thousands and millions of networks, all connected together: home networks and multinational corporate networks, supercomputers, and programmable light switches. Instead of predictable and stable patterns, a few people with a wacky idea will come out of nowhere, take the Internet community by storm, and generate traffic patterns no one ever predicted.

The difference between the official proposal—it was called "X.25," by the way—and the Internet proposal is that the Internet proposal was based on disaggregation and the X.25 proposal was not, . . . well, sort of. The X.25 proposal and the Internet proposal both included some of the same fundamental disaggregations. It's what *else* they decided to disaggregate that made the difference.

The Ancient History of Networks: The 1970s and 1980s

In 1987 I joined Bell Labs, which at the time was the world's greatest industrial laboratory. Bell Labs was a famous center for computer research—home of such remarkable innovations as the C programming language and the UNIX operating system. But when I wanted to send a file from my desktop computer to a colleague's desktop computer I used "sneakernet": I put the file on a disk, tied on my sneakers, walked to the colleague's office, and handed it over directly—even though we were both connected to computer networks. Why not just send files over the network? The problem was that even at Bell Labs, a center for network research, computers didn't connect very well. Or, in the language of disaggregation, the computers didn't have *interfaces* to help them interoperate properly.

The lack of interfaces points to something subtle: even though these were separate computer systems, they were not disaggregated computer systems. Here's what I mean. The first computers were huge "mainframes," and people programmed on them directly. Manufacturers eventually figured out how to connect mainframes with each other to create networks, but for the most part they never bothered to connect mainframes from *different manufacturers* into a single network. Inter-network connections simply weren't on the agenda; the manufacturers wanted to lock their customers into their proprietary computers and networks, and certainly didn't want to encourage their customers to wander off and hook some other manufacturer's computer to their network. When desktop computers came along, they were an independent invention. Networks eventually included desktop computers, but again in a very ad-hoc fashion.

Because the computers didn't disaggregate from a nice unified network, the manufacturers didn't go through the steps outlined in Chapter 4—in particular, they didn't build interfaces to keep the unified network up and running because there was no unified network in the first place. Manufacturers developed their own networks as they developed their computers, and people struggled after the fact to get these networks connected with each other. It was a very difficult task, for both of technical reasons (e.g., lack of software) and social reasons (e.g., patents). It was hard enough to get a network running at a single building at Bell Labs; if you wanted to connect a computer in Chicago to a computer in New Jersey, it was even more difficult; essentially you had to build a custom connection over the telecommunications network.

Two Visions of the Worldwide Data Network

Businesses continued to build data networks, and the inefficiencies of the existing system of ad-hoc network connections started to annoy everyone. In the 1970s, the major telecommunications companies decided that they would build data networks—after all, they owned almost all the wires anyway. The decided to build a data network that worked like the telephone network; big customers would get their own special connections and everyone else would dial in. To build a network that operated across national boundaries, the telecommunications companies needed well-defined interfaces, which in turn meant they needed to set standards. The International Telecommunications Union (ITU) is the telecommunications industry's main standards body; it's been around since 1865 and is currently the standards forum "blessed" by the United Nations. A committee of the ITU eventually produced the X.25 specification for data networks.

As might be expected from a specification produced in the 1970s and 1980s, X.25 assumed intrinsically that everything would remain the same as it had always been. Countries would have just one large, state-owned telecommunications company (except for the United States). Computers would remain big and expensive—ordinary people would dial into mainframe computers from home terminals if they needed a computer.

The academic community worked just as hard as the business community to network computers together. The academics worked harder, in fact; a business could just purchase all their equipment from one manufacturer, but the scientific community had many highly specialized computers from many different manufacturers, and they wanted to share not just data but actual computer time and resources. This was a much more difficult task.

Each university had its own internal network, and the goal of the universities was to tie these networks together. The academics weren't interested in building One Big Data Network; their "Internet" proposal focused on the question of moving data *between* networks, whether they were small networks confined to a single room or giant networks that reached across the entire world. There could be as many networks as people cared to build. The academic community developed a set of standards and defined the interfaces that make up the Internet.

Now there were two proposals on the table, X.25 and Internet. Both were standards (disaggregations of authority), but from very different groups. The big telecommunications companies—and by implication the governments that owned most of these companies—promoted the X.25 specification (to the extent they were interested in data networking at all). These big telecommunications companies owned all the wires, had experience in delivering high-quality service, and had contacts and business deals already in place with the large companies that would be the first customers. Equipment manufacturers were eager to build whatever the telecommunications companies wanted to buy. The academic team, with their Internet proposal, had a limited budget and limited clout. They didn't even own any wires to transfer data—when they needed bandwidth, they rented it from the telecommunications companies!

It's obvious that X.25 should have come out on top, and it's a bit startling that the Internet won instead. But after a closer look at the details of how the Internet works—it's even *more* startling.

How and Why the Internet Won

Why did the Internet proposal win? The answer lies in the details. The Internet deliberately disaggregated their solution at multiple levels. In fact, there's a great deal of similarity between the railroads and the telecoms, on the one hand, and the automobile and the Internet, on the other hand.

The Roadway

Like the railroads, the big telecommunications companies planned very expensive "roadways" for their data networks.

To explain how the telecom industry builds things, I'm going to relate a little-known chapter in a well-known tragedy. On September 11, 2001, terrorists destroyed the two main towers of the World Trade Center, and their collapse damaged other buildings in the World Trade

Center complex. Building 7 subsequently collapsed, and, as it fell, it severely damaged a nearby building that belongs to a local telecommunications company. This building housed the "switches" that route telephone calls. Falling debris ripped a hole in the facade ten stories high; some of the switches were blown halfway out of the building windows. The fire department ran hoses through the telecom building to get at nearby fires, which got the switches nice and wet; and the basement, filled with cables that connected the switches to the phone lines, completely flooded. Despite all this, the switches continued to work and calls went through. And another telephone switch *in the basement* of one of the collapsed World Trade Center towers was still working the next day! *That's* how telecommunications companies build their hardware—hideously expensive, "gold plated," but very, very, *very* reliable.

That story illustrates the mindset that the telecom companies had when they designed the data network. They set out to create a network that would transmit data perfectly. Since moving data along wires always introduces errors, they created elaborate error correction protocols to clean up the data—they wanted to guarantee that when data that entered the network at one end, it would be exactly the same when it came out the other end. After all, if the data couldn't be guaranteed as reliable, what would be the point in building the network in the first place? Their design called for expensive "gateways" to run the error-checking routines and make certain that the data were clean and perfect.

The Internet engineers took a different approach. Early academic networks, such as the famous ARPANET, were built in part to develop data networks for military applications. The military doesn't have the luxury of assuming networks will always be perfectly engineered—in fact, they have to assume networks will be noisy or damaged. The Internet engineers took their experience with less-than-perfect networks and decided to include them in their design. To make this work, they took a very important step: they disaggregated the task of making certain the data were perfect from the task of carrying the data. The network would carry data; handling errors would be a separate task.

The task of error detection and correction went to the computers at either end of the network—the ones that were sending each other data—based on two theories. The first theory was that, although it was a bit of a stretch at the time, computers would eventually be able to handle error detection and correction effortlessly because computing power would become less and less expensive.

The second theory was that network bandwidth would be plentiful and cheap, and you could "waste" bandwidth by retransmitting messages instead of getting them right in the first place. With error correction handled by the computers at either end, and with no worries about the wasted bandwidth, gateways between networks would just pass data between the networks; Internet gateways would be far less expensive and far less complex than the error-checking X.25 gateways.

They also decided on another idea that gave telecom engineers fits: if the Internet gateways were overloaded, the gateways could "drop packets on the floor"—discard the overflow of data—and the computers at either end would resolve any problems caused by missing packets. This made Internet gateways even less expensive, because they didn't have to store packets if traffic overloaded—instead, it was "move 'em or lose 'em." *Throw data away?* This was rank heresy to telecom engineers, and absolutely contrary to their notion of a well-engineered network.

So, what are our choices here? On the one hand, we have X.25, which guarantees delivery of clean data; and other the hand, we the Internet, which seems to guarantee the exact opposite. The choice would seem to be pretty clear—go with the clean data and X.25. What advantages could the Internet possibly offer?

The Internet did offer advantages, decisive ones. Internet gateways—we call them *routers* now—were affordable. Anyone could join the Internet; you didn't need to be a huge telecommunications company; all you needed was a connection, a gateway, and the right software on your computer. You didn't have to meet rigorous engineering standards; if your network was noisy or lost a lot of packets, that was your problem.

But was the Internet really practical? After all, can a network actually work if the data can be corrupt or disappear? A good day for me to answer that question was the very day I typed these words into the typescript for this book: I was waiting for my Internet service provider to show up and fix my office's connection to the Internet—up to half of the packets I was sending and receiving were getting lost. But even with half the packets missing the connection still generally worked; when data disappears, the computers on either end of an Internet connection retransmit the data. What I experienced was a slow and quirky connection—not one that had failed entirely. Now don't get me wrong—when a connection is running slowly and losing data, people won't stand for it (which is why I impatiently waited for repair). But that's not the same thing as saying that the network has to keep the data nice and clean at all times; even when there's a failure, the connection can remain somewhat useable. For all its faults, my office connection is far less expensive than a perfectly engineered connection, the kind that X.25 requires. And the ability to error-correct means that if the connection does go bad, it can still be usable.

The Internet's acceptance of less-than-perfect, its disaggregation of data integrity from data networking, encourages creativity. If I invent a service where lost data is perfectly acceptable, then I can go ahead and create that service. (The typical example of that kind of service is broadcast video over the Internet—it's OK to lose a few packets of video information now and then.) The basic Internet philosophy lets *participants* decide on how much reliability they need for their network, rather than dictate perfection in all things.

The Traffic

As previously mentioned, the X.25 network used error correction in its gateways. This means that the traffic on its network—the packets of data traveling on it—had to be data the gateways would recognize; otherwise, the gateways wouldn't be able to differentiate between errors and legitimate traffic. To make gateways work, X.25 placed restrictions on traffic and allowed only certain types of data packets on the network.

Because the Internet disaggregates the job of verifying the data from the job of passing the data between computers, the Internet is much more flexible. The Internet specifies a kind of "envelope" for data packets. What's inside that envelope isn't something the Internet needs to know about; that is, although it's necessary for the *computers* on either end to know what's in the envelope, the *network* does not need to know. Anyone can send any kind of content they like, as long as they stick the data into properly marked envelopes—and they are allowed to create new kinds of envelopes. These lax rules on data traffic encourage a tremendous, continuous outpouring of innovation as people use the flexibility of Internet specifications to experiment with transmitting all sorts of data.

Network Management

After deciding how to build the roadbed and the rails, the railroads decided where the rails would lead, what kind of equipment would run on the rails, and how the trains would travel on the rails, and they devised timetables to control when and where the trains would run.

The X.25 system relied on very similar management; X.25 networks would connect to computers at large companies, universities, and similar institutions. Everyone else would use terminals to log onto mainframe computers. Connections would be managed as a scarce resource (and would no doubt remain both scarce and expensive). In theory, the X.25 "address space"—the range of permissible network addresses—could support more than one data network per country, but outside of the United States that didn't seem likely, and even the United States was assigned only one hundred networks. The new data network would be very much like the telephone network, with every aspect tightly controlled by central management.

The Internet proposal let anyone join the network; because the Internet engineers didn't demand that networks meet any particular level of quality, effectively there were no restrictions. In contrast to the X.25 plan, the Internet "address space," the range of permissible network addresses, allowed for literally millions of different networks.

As discussed earlier, another important aspect of running a network is "routing"—determining which wires to use to send a data packet from Computer A to Computer B. Let's say I work in the post office and I need to route a letter. I can look at the city name, the street name, and the house number, and that'll give me a pretty good idea of what streets to take to deliver the letter. But what about a data packet that your computer sends my computer—how do I route that data? What wires do I put that data packet on? The address on the data doesn't say "Moshe's office network in Chicago"; it doesn't even say "the disaggregate.com network." Instead it has a computer-readable address—for example, 192.0.34.166—and there's no clear relationship between this number and where my office is. It's not any sort of geographic code; it's just an arbitrary string of numbers. How does data from your computer find its way to my computer?

Routing under X.25 was designed to work in the same way as routing did for railroads and the telephone system: with reserved paths and central management. To send data from Computer A to Computer B, the network would determine the exact endpoints of the data call and what wires to use, and would then allocate a path through the network for the data. This style of top-down routing requires extensive management resources to control, reserve, and monitor the network.

By contrast, the Internet engineers deliberately disaggregated the authority to determine paths through the network from any central authority and gave it to the local networks to sort out. Data on the Internet acts like automobiles: the data is divided into packets, and each little packet of data finds its own way through the network. Here's what I mean. When Computer A sends a message to Computer B, Computer A doesn't chart a path through the network to Computer B. Instead, Computer A breaks its message down into packets and sends them off, one after another, addressed to B. Packets go down the wire until they reach a router. The router decides where to send the packet next by looking at the source (Computer A) and the destination (Computer B), and applying any other criteria that the router might have.

Let's say the router gets the very first packet. It picks a route and sends the packet on its way. Then the router gets the second packet, but a quick check shows that the previous route is now congested. In that case, the router might send this next packet, which is going to the very same destination, along a different path. The router hasn't reserved any resources for exclusive use of the conversation between A and B; it simply pushes packets around in an attempt to keep traffic moving and use local resources as effectively as possible. The packets move through the network this way, getting closer and closer to Computer B until they finally arrive. Sometimes packets arrive out of order, and sometimes a packet will be discarded if a router is too congested. It's up to the computers at either end to sort out what arrived, what didn't, and what came out of order; the Internet protocols specify how to resolve any problems.

This harum-scarum delivery system for data is what makes the Internet cheap and reliable. By forgoing centrally managed paths, local routing decisions are disaggregated from routing decisions at the national or international level. There isn't a "National Internet Traffic Control Center" to micromanage every decision in the Internet any more than there's a national agency to tell automobiles how to drive from Point A to Point B.

The benefit is simplicity coupled with enormous cost savings. Planning on a nationwide scale—replicating the timetables of the railroad days—doesn't have to happen; every one of the millions of networks in the country decides for itself how to route its traffic. Routers for small local networks can be inexpensive and "dumb" because they only confront a limited set of problems. A network that relies on local decisions is also highly reliable. If a construction worker snaps a data cable with his backhoe, local routers notice that the link is down and send the packets somewhere else, automatically—the process of "healing" the network happens at the local level. The network may not be very efficient in terms of how many miles a packet travels to get to its destination, but other efficiencies are overwhelming.

The Business

The Internet's original design did not include disaggregations that were aimed specifically at business issues. The innovations and choices at other levels—in particular, the acceptance of millions of networks—meant that the business infrastructure of the Internet would be sufficiently chaotic without any further intervention. Instead, I'll discuss how the Internet's disaggregations resulted in business versions of the universal benefits; the examples will be competition, cost reduction, and specialization.

The ability of people to form their own networks and connect them together using Internet interfaces means that no one company can, for technical reasons alone, control or dominate the business of providing Internet connectivity. This doesn't mean that companies won't apply social measures to attempt to dominate the Internet business, and some have; but in general there's good competition. For example, as the Internet started to take off, some Internet service providers attempted to impose the business model used by telephony companies: they charged users by the byte in the same way telephone companies charge callers by the minute. This fell to the wayside (for the most part) as competitors sprung up with less-annoying business models. The proliferation of competitors also drove down the cost of service.

Another business implication is the rise of specialized networks as people find business niches and fill them. Cellular telephones now have Internet addresses and offer a range of Internet services, including web browsing, instant messages, and e-mail; entrepreneurs are installing wireless networks for business travelers; other businesses and business models come and go on a regular basis. Some networks are optimized for browsing, others are optimized for video. Some are optimized for short messages, and others are ad-hoc wireless networks cobbled together to provide free Internet access. The Internet's openness encourages the growth of these highly specialized networks and others besides.

The Internet model also helps prevent attempts by large corporations to censor the network. If X.25 had triumphed and the international data network was managed on a "one large provider per country" basis, vested interests could stomp on ideas they don't like. Take file sharing, for example; some businesses hate the idea of file sharing, and would love nothing more than to serve an injunction on a single company and get file sharing stopped forever. Today's fragmented networks of the Internet all have different policies, management, and legal jurisdictions; this makes it hard for an industry to throw its weight around in attempts to suppress new technologies. If the data networks were all under the jurisdiction of a few companies and those companies were publicly regulated utilities, it's easy to see how a new way to share data—or maybe even the data itself—would be subject to political review and control.

Table 7.1 summarizes information from Chapters 6 and 7. The table compares disaggregation in the transportation network and the data networks. The railroad, which was the dominant technology, and X.25, which was proposed by the dominant technology companies, were overturned by the disaggregations introduced by the automobile and the Internet. Note how the transportation networks and data networks have similar infrastructure layers.

Table 7.1 Comparison of Pre- and Post-Disaggregation Networks (Transportation and Data)

Railroad	X.25		Automobile	Internet
Trains run on carefully engineered, very expensive tracks.	Network connections must be highly reliable to prevent data loss.	ROADWAY	Automobiles move on any road surface. Roads can be anywhere, inexpensive.	Internet traffic travels over any network, even noisy ones. Error correction is disaggregated from the job of carrying data.
Traffic on the railroad must fit on tracks, pass through railway tunnels, and be compatible with rail yard equipment.	Data traffic must be recognized by the network so the network can verify its integrity.	TRAFFIC	Dimensions are not rigidly defined by need to move on tracks; includes everything from motorcycles to tractor-trailer trucks.	Only minimal requirements for allowing data to move through the network; no restrictions on types of data.
Tracks are a scarce resource.Cargo is the responsibility of the railroad. meticulous central planning keeps cargo moving and trains running.	Bandwidth is a scarce resource. Network managers conserve these resources and regulate traffic flow.	NETWORK MANAGEMENT	No need for master schedules. Autos go where they please, when they please.	Each separate network manages internal affairs. Decisions on how to get data to its destination are made locally at each router.
Railroads control the business of transportation over their rails.	Networks control the business of data flow over their networks.	BUSINESS	Any company can build service based on automobile; no need for central management approval.	The Internet consists of millions of networks; plenty of opportunity for different business models.

Benefits of the Permanent Revolution

One of the most important legacies of the invention of the Internet is the associated attitude of permanent revolution. Back in Chapter 4 we discussed the final stage of disaggregation, namely, "evaluate." After an innovation starts a revolution, it's worthwhile to evaluate what happened, the benefits of the innovation, and what should happen next; the information can then be used to launch the next round of disaggregation. The Internet community absolutely embraces this philosophy.

The Internet engineers fostered a culture that constantly disaggregates interfaces into smaller and smaller building blocks. The interfaces are generally created in small pieces in the first place rather than as one grand masterpiece; and once an interface has been out for a while, Internet engineers tend to disaggregate it into even smaller pieces. The result of this mindset is a continuous and spectacular avalanche of new ideas on the Internet.

To show how this works in practice, I have examples that illustrate synergy, simplicity, and specialization. I will start with an example of simplicity: how the Internet creates interfaces that are composed of simple parts, and how that simplicity helps the Internet keep its interfaces up to date.

E-mail wasn't part of the original Internet proposals. Although it already existed in some non-Internet systems, no one realized how popular it would become, and it simply wasn't covered in the original blitz of proposals. Internet-based e-mail started out as a "hack" put together out of existing systems. Because e-mail hadn't been standardized, various people came up with the idea at different times, and pretty soon there were several incompatible e-mail message systems running on the Internet. Eventually, Internet engineers got together to develop standard ways to send and receive e-mail across the entire network.

How does e-mail work? It's pretty simple on the surface. I type in a message on my computer using a program like Eudora or Thunderbird, I put your address into the mail, and when I'm finished I push the "Send" button. Sometime later you check your e-mail and my message arrives at your computer. What's happening behind the scenes is this: my computer talks to a computer at my e-mail service provider (usually it's just the Internet service provider) and places the message into the e-mail system, and your computer talks to your e-mail service provider's computers to retrieve the message. The Internet engineers had to create an interface that all the computers could use to pass the messages around.

At first glance, writing an e-mail interface for the computers to use to send e-mail to each other ought to be simple enough: here's how you send, here's how you receive. But the Internet engineers chose, very wisely, to disaggregate the interface for e-mail into many smaller interfaces. There are interfaces for:

- Sending e-mail.
- Retrieving e-mail. There are two main interfaces in use, and each offers its own advantages.
- An interface to explain how messages are formatted—the correct way to write the "To" and "From" addresses, what's allowed and not allowed in the "Subject" line, and similar requirements.
- An interface to explain how to send attachments in e-mail.

Because all of these tasks are handled separately, the interfaces are far simpler, can evolve more rapidly, and are easier to maintain. For example, take the interface for sending e-mail. Recently some people and organizations decided to enhance security to help cut back a little on spam. After some sparring, they've come up with a new version of the interface—but they only had to deal with *that single interface.* Not only is that a simpler technical change, it's a simpler social dynamic. They didn't have to coordinate their changes with the people who maintain the other e-mail interfaces—they didn't have to horse-trade changes in one area for changes in another, they didn't have to wait until everyone was ready to publish their revisions, and so on. The problem could be solved with just the interested parties working on the deal.

Synergy is another important benefit of disaggregation. The Internet's focus on continuous disaggregation has created many, many smaller and more flexible interfaces that just beg to be used to create new services—and new businesses—on the Internet. The World Wide Web relies on the *hypertext transfer protocol*, which was invented to move Web pages around (that's the "http://" you see in Web site addresses). But now all sorts of other Internet services incorporate the convenient, well-supported, and very handy HTTP interface to move other sorts of data around. Like other tools on the Internet, no one cares what it was invented for if it solves the problem, and HTTP turns out to be a very handy tool indeed.

Finally, here's an example of specialization. *Internet telephony*, making telephone calls over the Internet instead of over the classical telephone network, is rapidly gaining acceptance and in a few years will displace the classical network. There's an interesting twist to Internet telephony: the interface that defines how to carry the audio portion of the call doesn't attempt to dictate how the audio is compressed. That is to say, the two interfaces—the "here's the audio" interface and the "this is how to listen to it" interface—are disaggregated from each other. From time to time, people suggest clever new ways to compress audio, or clever new things to put into the audio portion of the call that aren't really audio. It's a highly specialized interface that lets experts be very creative.

The Internet was founded on a series of disaggregations, and the tradition of disaggregation continues. It's this tradition, this permanent revolution, that explains why the Internet continues to transform business, technology, and society.

The Automobile and the Internet

The automobile focused on a single disaggregation, moving mechanical transport off the railways and onto the ground. The Internet is a continuing series of disaggregations, a permanent revolution. In the end, both the automobile and the Internet accomplished the same goal, the disaggregation of an industry at multiple levels.

As always, these revolutions didn't happen by having people issue manifestos and then sit around waiting for the revolution to start. Both revolutions took hard work. Automobiles had to be sold to an initially skeptical public and faced a great deal of outright hostility; it wasn't obvious to everyone at the time what a fantastic idea the automobile was. The same holds true for the Internet. Inventors didn't just design a better mousetrap; they worked for years to get it adopted. Today's innovations often face the same problems of indifference and hostility, which means hard work to get an innovation adopted.

But don't be discouraged. Henry Ford made a fortune promoting the automobile, using the techniques of mass production. The pioneers of the Internet era did very nicely for themselves—and I'm not speaking of the Internet bubble in the late 1990s, but of solid companies such as 3Com and Cisco. Both sets of innovators understood that a revolution was in progress, and they profited from it. Spotting the right innovation early can be very rewarding indeed.

Business Strategies: How to Cope, How to Fail, and How to Predict the Future

Part III

Part III includes information on business topics.

Explains standards and how they fit into business strategies.

Discusses how to cope with avalanches in your industry—what to do when you're suddenly confronted by a revolutionary innovation.

Illustrates how **not** to cope with avalanches in your industry, based on examples of spectacular failure. The chapter also discusses how counterrevolutionaries attack innovations.

Raises the problem of coping with government intervention.

Analyzes three industries poised on the brink of revolutionary change.

Provides a roadmap for putting the ideas of this book into practice, and it includes a few final thoughts about what's in the book—and what comes next.

Chapter Eight

Interfaces and Standards: The Nuts and Bolts of Modern Civilization

Somewhere in my basement I have my Ph.D. thesis in electronic format on a couple of disks—eight-inch floppy disks, to be exact. Floppy disks that size haven't been made for years, and they're about as useful with my present computer as a pair of stone tablets. Well, actually, stone tablets would be *more* useful; I could *read* stone tablets. I'll never be able to read those old floppy disks.

My basement is full of old computer equipment that still works but isn't useful anymore—the equipment is just not compatible with modern computers. I've got both dot-matrix and daisy-wheel printers, a brand-new six-pen plotter along with dried-out pens, a stack of Apple's old Mac Plus computers, and a defective TRS-80 Model 100 laptop computer (16 kilobytes of RAM! Three-line LCD screen!). I've got old memory cards, a scanner without software, the guts of obsolete

Palm Pilots, and a tangled nest of cables that won't connect to anything. I often tell myself, or my wife, that I hold on to these old parts because they occasionally come in useful repairing old pieces of equipment. That does happen from time to time, but mostly it's just the pack rat instinct common to members of the technical community.

I think that those piles of obsolete equipment, some of it only a few years old, symbolize the way most people think about technology today. "Computers change so fast that they're obsolete before you make it home from the store." "None of this new equipment works with my old stuff." "I've got to replace my computer every few years or it stops working." "The old printer doesn't work with my new system."

Many people think that instant obsolescence is an inevitable consequence of progress. But they're wrong: immediate obsolescence is an aberration, not the norm. Otherwise, modern civilization would be utterly impossible.

Here's an example of how technology really works in our modern society, a demonstration of compatibility that really astonished me. Many years ago I lived in an apartment, and one day the ceiling light in the kitchen wouldn't go on—I pulled the chain but it wouldn't budge. I detached the fixture from the ceiling, opened it up, and discovered that the fixture dated from the late 1920s or early 1930s. The switch was broken and I decided I'd try to replace it with a new one, so I headed over to the local hardware store. To my surprise I was able to find a replacement switch that fit exactly into the old fixture—not only that, but the replacement was the same inexpensive standard switch made for modern fixtures! Even after fifty years, pull-chain switches and light fixtures still fit each other. This was absolutely *amazing*. At my laboratory I was struggling to get computer equipment that was only a few years old hooked into my physics experiment, but at home I had fifty-year old fixtures that I could repair in a snap!

When I thought about it some more, I realized there was something else odd about the light fixture—the light bulbs. A fifty-year-old piece of equipment, and yet I could still use modern light bulbs? That provided an odd contrast to my computer equipment, which I especially noticed on days when I would try to get a five-year old computer to interoperate with my new home network.

Actually, aside from computers, most of the stuff of day-to-day living has standard shapes and sizes as well as interoperable parts, and doesn't change very much if at all. In fact, modern civilization depends on it: we keep our civilization running with standardized parts. Plumbing can be fixed because major valves, fittings, and pipe sizes are all standardized; imagine the chaos if each plumbing company made its own sizes. Home exteriors can be repaired because masonry bricks and cinder blocks come in the same sizes they always have. Light bulbs fit into old fixtures, modern appliances plug into old electrical sockets, and the electricity supplied by the power company is the same voltage and frequency all across the country. Gasoline for automobiles can be purchased at any service station—the auto manufacturers do not require special brands of gasoline made only for their vehicles. Basketballs fit through basketball hoops, photographs fit into picture frames, and clothes hangers fit onto the hanger bar in your closet.

So on the one hand our modern life depends on interoperable parts in standard shapes and sizes, and on the other hand some of our crucial modern tools (e.g., computers) have parts that seem to barely interoperate and can't be salvaged if they're a few years old. That's pretty confusing, and it raises a series of interesting questions:

- Why are standard parts so widespread throughout modern society?
- When did people start to make standardized parts, and why?
- What is the business case for standards?
- What are the business penalties for not using standards?
- Why is it easier to fix a fifty-year-old light fixture than a two-year-old computer? Have standards failed when it comes to computers?

The Origin of Standard Parts

The people who introduced the idea of standardization for all the "nuts and bolts" of modern civilization were, ironically, the manufacturers of nuts and bolts. Let's leave computers aside and return to the early days of mass production as described in Chapter 5—the early 1800s, the days of Whitney and Terry, when the ideas of mass production first took hold.

Just about every mechanical product (both then and now) uses nuts and bolts: sewing machines, doorknobs, boilers, printing presses, stagecoaches—you name it and it has nuts and bolts. Disaggregation breeds specialization, so it was only natural that some manufacturers began to specialize in making nuts and bolts for other manufacturers. This *machine screw* industry, as it's often known, was (and still is) a competitive industry that produced high-quality components at reduced cost. But there was one fly in the ointment: each manufacturer of nuts and bolts had its own notion of what dimensions to use for its product.

The dimensions of the nuts and bolts—the diameters, the lengths, how far apart and what shape the threads are—determine whether a nut and bolt will fit together, but the manufacturers had no incentive to interface with each other. This wasn't a specialized industry like the wooden clock industry, in which all the wooden gears had to fit together or the clock wouldn't work. If I made doorknobs and you made sewing machines, there was no reason why a bolt from my doorknob had to fit a nut from your sewing machine. Neither of us would ask our respective suppliers for identical nuts and bolts; without customer demand, the manufacturers didn't *have* to go to the trouble of making nuts and bolts that interoperated—and so, of course, they didn't. In the early 1800's there was no reason to expect that a nut or bolt from one nut and bolt manufacturer would fit those from another manufacturer, and in fact every reason to expect that they did not.

Incompatible nuts and bolts made repairs difficult, to say the least. Today I've got a chest of nuts and bolts downstairs, right next to that stack of computer parts, and when I have to fix something around the house I can reach into a drawer and pull out the nut and bolt I need. If there's a size that I don't have, I wander off to my hardware store and pick it up for a few pennies. But if every time I needed to replace a nut or bolt I had to contact the product manufacturer, or worse yet find the specific company that made the nut and bolt, two things would happen: my repair costs would shoot through the roof, and the time to make a repair would increase from minutes to days. (If you live in the United States and you've tried to fix a bicycle that uses metric nuts and bolts, you've probably had a taste of this problem.)

There's also cost to the person who designs and manufacturers the widget that uses nuts and bolts. Let's say I manufacture doorknobs. I can't just go out and get standard tools to use on my production line to drill holes for the bolts; if each manufacturer has different size bolts, there won't be any standard tools. I have to make the tools myself (expensive) or get them custom-made (also expensive). If I suddenly need to use nuts and bolts from a different supplier—because my old supplier went out of business, increased his prices, or started sending me low-quality parts—I'm in trouble. Nuts and bolts from a different supplier aren't likely to fit my current product or production line. And that also makes it costly to create new products, such as a fancier doorknob, because I'll likely have to purchase new tools instead of just using the tools I already have. So it wasn't only the customers who suffered from incompatible nuts and bolts; manufacturers suffered too.

As always, people could get by. If someone lived in a little town somewhere, they'd buy their doorknobs in that little town, and the repair parts would probably also be available locally; or if not, the local blacksmith could make something that fit. Doorknobs are pretty simple machines; the problem got worse as the machines became more complex, which in turn made people a little wary of purchasing anything that used a lot of nuts and bolts.

In Chapter 5, we discussed how in the early 1800s people didn't have many mechanical objects in their homes. Remember how wooden clocks sold for $40 in 1800? In today's money, that's about $600! You could buy forty acres of property in New York state, and a house, for just $50 in 1820. The costs of nuts and bolts made the little machines we take for granted today horrifically expensive back then. Because the little machines were expensive, not many were sold, and low sales meant that the primary market for nuts and bolts couldn't grow.

For one high-tech industry, the repair problem was nearly impossible. The important, exciting high-technology industry of the early 1800s was the railroad industry, with steam locomotives and railroad cars that traveled all across the country. The railroads had repair depots, of course, but the depots couldn't rely on local suppliers for the "simple" nuts and bolts that held their machines together. Railroad depots could stock their nuts and bolts directly from the factory, although that was expensive. But if a locomotive broke down in a small town right next to the local hardware store and couldn't be repaired using the local varieties of nuts and bolts—that was ludicrous.

In the mid-1800s, first the English and then the American nut and bolt manufacturers decided to do something about these problems—the high repair and manufacturing costs, the lack of market growth, and the complaints from high-tech industries. The Franklin Institute in the United States, a professional organization for machinists, introduced voluntary national standards for the dimensions of nuts and bolts (the Americans didn't adopt the English standard for technical reasons). In other words, the manufacturers disaggregated their authority to determine the dimensions of nuts and bolts. The standards acted as an interface to let the manufacturers cooperate and communicate—they were no longer isolated from each other.

The changeover from the chaos of different shapes and sizes to the new standard ones was emphatically *not* simple, easy, or straightforward. It was very different from the natural changeover to Whitney and Terry's methods of mass production that we saw back in Chapter 5. Mass production did reduce costs; a clockmaker could cheaply produce thousands of clocks a year. But when a manufacturer adopts a standard, it *costs* money: adapting or replacing machinery, changing product lines, checking products to make certain they meet the standards. All that work, and for what benefit? The manufacturer still ends up making nuts and bolts, just in slightly different shapes. Sure the railroads were a big customer, but if you had a company that made nuts and bolts, and every single nut and bolt that you sold went to my factory to make doorknobs, why would you care about what the railroads wanted? And let's not forget me, the doorknob manufacturer—I'd have to spend lots of money to change *my* entire production line to handle the new sizes of nuts and bolts, and I guarantee you that I wouldn't do it just to accommodate the railroad barons!

The Business Case for Standards

A national standard only works if everyone switches over; if only a few companies adopt the standard, there isn't much point. The revolutionaries who introduced the innovation had to present a solid business case or the idea would never fly.

In the case of nuts and bolts, the manufacturers had a long debate about the merits of standards. There's good news and bad news about the debate. The good news is that the industry eventually adopted the standard. The bad news is that, as I read through the history of the discussions, I recognized *every single* debating point, because 150 years later when I led an industry standards group on speech technology we debated the exact same issues—not exactly what I call progress!

To examine the arguments, let's assume that we own a company that makes components, which we sell to customers to build into their widgets. If we propose that our company reengineer our components to meet a new industry standard, two groups within our company are likely to step forward to object: the first group is the sales and marketing teams, and the second group is the engineering team.

Let's give sales and marketing first crack at persuading us not to use this new standard. The first argument we'd hear is that any move to eliminate differences also eliminates a competitive advantage. Think about our company's customers. Once our components are designed into a customer's widgets, we've got a terrific competitive advantage because any attempt by the customer to switch to one of our competitors' components will be very expensive for the customer. To switch, customers must rewrite their software, redesign and rebuild their hardware, and teach their repair technicians and salespeople about the new component—all expensive undertakings. The cost of making a switch to a competitor helps "lock in" our customers, but that cost is far lower if we introduce standards.

Worse yet, what happens if our company plunges ahead and makes the new standards-based component? If a customer decides to use the new standardized component, I can guarantee that the customer will take that opportunity to think about switching to one of our competitors as long as the customer is paying for reengineering anyway (I can guarantee it because I've done exactly that myself). If we introduce a standard into our components, the argument goes, we put our customers "up for grabs."

And there are other arguments. Maybe our customers won't buy these new standard components until the dust settles on the new standard. Standards don't always succeed in the marketplace; after all, standards take time before they function smoothly while engineers work the bugs out, and some standards fail to deliver the benefits they promise. We might go through all the trouble of creating a standardized component and then have trouble finding a customer for it. Maybe we should be opposing this new standard instead of supporting it.

Even if customers do decide to adopt the standardized component, customers might defer their changeover to the new component as long as possible in order to defer their reengineering costs. These delays pose two problems. The first is that delays make it harder for us to recover our investment in the new component. The second is that delays and lost sales are personal nightmares to sales and marketing staff, who have to sell in order to earn their salaries (well, and earn enough for the company to cover the salaries of the engineering staff, too, but engineers don't usually think in those terms). And speaking of money, if there's a standard, the component might turn into a commodity—no surprise, because that's a benefit of disaggregation—and commodities generally have lower profit margins, which is something else the sales staff doesn't appreciate.

After the sales and marketing teams are through, the engineers will step in with their own series of objections. Some of their objections will likely boil down to that famous engineering malady known as NIH, which stands for "Not Invented Here." Engineers take pride in their work, just like anyone else, and they're reluctant to toss their designs aside and adopt someone else's just because it's a "standard." (And for that matter, the sales and marketing teams might not be happy if they've been touting the company's proprietary design as superior to everyone else's.) At other times, it's completely correct to oppose standards on technical grounds. Sometimes the standard is ill conceived, won't deliver as promised, or doesn't seem to have much in the way of industry support. This decision requires engineering judgment, tempered with management judgment.

Another problem engineers have with standards is that engineers like to tinker; and if you standardize something it can be the end of tinkering. Engineers will also present a budget for the project; meeting the standard will take time and money.

With opposition from the sales and marketing teams and the technical staff, you need some pretty convincing arguments to get companies to adopt a standard. In fact, you need some very convincing arguments because many of these objections from sales, marketing, and engineering are absolutely valid. Customers can switch to a competitor, and some will; your product may be commoditized; you will have to work harder to distinguish your product in the marketplace; and the technical staff will have to slow down their tinkering with improvements and instead spend time working with (and around) the standard to keep things running. To top it all off, standards can cost a lot of money to implement. With all these great arguments against standards, why are they ever adopted?

The answer is straightforward: adopting an industry standard can make excellent business sense. The nut and bolt industry, by setting standards, solved the problems that prevented people from using nuts and bolts in inexpensive machines:

Manufacture	Manufacturers could use standard tools—drill bits, taps, and the like—to make their products. The engineers didn't have to waste time learning about individual nuts and bolts; they could instead concentrate on how to use them to maximum advantage. Manufacturers weren't at the mercy of a single source for their nuts and bolts.
Repair	Since a nut from one manufacturer would fit a bolt from a different manufacturer, repairs became easier and less expensive, which made customers less reluctant to purchase widgets with lots of nuts and bolts.
Portability	With a national standard, a product built in one city could be repaired in a different city, which opened up the market for products on a nationwide scale. National standards also meant that a manufacturer of nuts and bolts could compete nationwide.

The result was that products assembled with nuts and bolts became less expensive, and, as they became less expensive, people bought more of them. This increased the size of the market for nuts and bolts, and it meant more business for the nut and bolt manufacturers.

Standards may mean new competitors and that you may lose market *share*, but making life easier for your customers grows your *total* market. "Maybe you'll have a smaller share of the pie," we used to explain to people who worried about competition, "but the pie will be much bigger." Would you rather have 20% of a $10-million industry or 10% of a $500-million industry?

▪

A good example of why standards succeed can be borrowed from the field of speech technology—the art of getting computers to respond to the human voice. If you want to let someone talk to a computer, you need two things. First, you need what's called the engine, which is jargon for technology that listens to people. Next you need dialogue software that talks to the user: it asks questions, turns the engine on and off to listen to the user's responses, and then decides what to do next. Building the engine is deep, dark magic, and only few companies in the world sell speech engines. But putting together a dialogue—using the engine to accomplish something useful—that's what the customers who buy the engines are supposed to do.

If speech magicians want to sell engines, customers have to write dialogue software. For many years, it was very difficult to write dialogue software. Most programmers never learned how to use speech engines at any time during their careers; the few who did, for the most part, learned how to use only one company's engine, and each and every engine was different. The speech industry never developed a pool of talented programmers who could write dialogues, and therefore business suffered.

Eventually the industry went down the same path as the nut and bolt manufacturers: they standardized the interfaces to the engines. This made it easier for programmers to learn how to use the engines and made it more attractive for programmers to enter the field because the skills they learned could be used more widely. The industry's pool of talented engineers continues to grow.

This new standard provides another important benefit: competition. Dialogues work with any engine that follows the standard. When a customer writes a dialogue—and good dialogues are very expensive to write—the customer can use it with engines from any vendor; there's no "lock in" to a single vendor. (At least in theory; the industry is still getting there as the standard becomes more mature.)

More programmers, simpler programming, and protection of customers' investments removed some of the barriers to the technology, and now even "little guys" use speech technology.

▪

The idea of standards was so successful that it began to spread throughout other industries. The example of the clock makers, with their informal interoperability, and the nut and bolt industry, with their formal standards, proved impossible to ignore. Today, literally hundreds of thousands of standards, ranging from single-industry informal agreements to UN-supported international agreements, support businesses in their march toward better interoperability.

Business Case Errors

Some industries are absolutely determined to shoot themselves in the foot. The data CD industry managed to carefully disaggregate the specification for data CDs from any particular manufacturer, which made the CD a very useful worldwide standard. The result: at the time of this writing, all new computers have a drive that can read data CDs and many have drives that can write them.

What about DVDs for computers? As their prices have dropped, DVD drives began to gradually replace CD-only drives. Because DVDs have enormous capacity, instead of writing seven or so CDs I can write one DVD. That'd be pretty useful for a computer setup like mine—but the DVD industry didn't unite behind a single standard for writing DVDs, which created all sorts of trouble. As I typed this manuscript, there was a battle in progress on the shelves of your local computer store: DVD-R versus DVD+R, two competing formats with differences that are important to the manufacturers but make little difference to the majority of people who purchase the drives.

Given the confusion, lots of ordinary people are sitting on the sidelines, holding onto their wallets and waiting for the dust to settle instead of upgrading their computer systems with DVDs (I'm one of the holdouts). Some manufacturers have thrown up their hands and started producing "DVD±R," that is, drives that support both formats simultaneously. It's like the great war a generation ago between Betamax and VHS videotapes; even some of the combatants are the same! Sure the stakes are high, but during the war the industry loses sales as people don't buy, and the fight alienates the customers who do buy but end up with the losing format. To cap it off, there's another DVD fight just starting. At the time of this writing, we expect the second generation of writeable DVDs to appear soon on store shelves—"blue laser" DVDs, which hold at least five times as much information as current DVDs—but *that* standard is in dispute as well. Don't hold your breath waiting for these new-style DVDs.

Given the clear advantages of a uniform format, and the disadvantages of a feud, why is it that the sides can't agree on a single format? The answer is that they're not just fighting over whose technology is better; there's something else at stake. They're fighting over something near and dear to their hearts: money. Both sides hold patents, and whoever wins the battle gets to collect money for those patents. This isn't an insoluble problem; the usual solution is to place the patents into a pool and share the revenues, a solution that dates from the 1850s. In the DVD dispute, each camp formed its own pool. Each side apparently believes that it has an excellent chance to win all the money rather than share any with the opposite side.

What went wrong? A single DVD standard requires two disaggregations: the disaggregation of authority (for the specification) and of ownership (patent rights). Because the sides can't agree to share, they can't disaggregate ownership. The lesson is twofold. First, although disaggregation offers potent advantages, it isn't magic and it can't always overcome human nature. Second, failure to come to a resolution or a single standard has very real consequences; in the case of DVDs, the slower adoption of DVDs translates to lower return on investment for both sides.

Standards and Obsolescence

What about my complaint at the beginning of this chapter, about computer equipment and how it becomes useless so quickly? I can understand why I can fix a fifty-year-old light switch with an off-the-shelf part—there's a standard of some sort for the switches. But what about computers? Computer standards don't persist for fifty years—fifty weeks seems more like it. What's the difference between light switches and computers? Why doesn't disaggregation—in this case, standards—affect them both in the same way?

There are two good reasons why light switches don't change over time, but computers do: 1) the fundamental technology and 2) the extent of disaggregation.

The fundamental technology of light switches, with all due and very sincere respect to the engineers who design them, is dirt simple: a piece of metal snaps back and forth. Remember those big "knife" switches—the ones that mad scientists use in old movies? Well, home light switches look like that on the inside. I've had occasion over the years to open up switches, and they're just not that complicated: a metal bar, some screws, and usually a spring. In the "on" state, switches carry a few amperes of current at 120 or 240 volts; in the "off" state, they don't, and that's about it. They're good, they're reliable—it's not easy to make a switch that lasts fifty years—but the technology doesn't change all that much.

Computer technology changes from year to year, from month to month, and even from day to day, and as it changes it improves drastically. So even though there's often nothing lost by staying with old-technology light switches, in computers newer technology brings advantages. For example, consider the floppy disk. You might remember 5 ¼-inch floppy disks; they could hold about 100 kilobytes when first introduced. Today, the floppy disk measures 3 ½ inches and holds fourteen times more data, has an integral plastic case to protect the disk against damage, can't be accidentally inserted into the drive upside down or sideways, and is small enough to fit into a shirt pocket. (And even so they're obsolete, replaced by other storage media such as the CD and the flash drive.)

The other reason for slow change is the *extent* of disaggregation. Light switches aren't as mechanically disaggregated as computers are. A switch that goes inside a fixture has to be the right size or it just won't fit; this means that for there to be any change, the manufacturers of light switches and the manufacturers of fixtures must both agree on the change, and there's a huge base of of already installed fixtures that use the prevailing standard. Absent any compelling reason—any *economic* reason—they won't bother with a new standard. The same holds true, maybe even more so, for light switches mounted on walls. Your wall light switch is actually screwed into a standard-sized box that's built into the wall. Let's say some manufacturer decides to build new light switches, big round ones or little cute triangular ones. If these new ones won't fit into the standard boxes that are in every home, how will customers install them? The manufacturer would have to convince homeowners to buy the new switches *and* tear the old boxes out of their walls *and* install new ones—but who would rip their walls apart just to change the shape of a light switch?

Computers don't have this problem because computers are more mechanically disaggregated: the computer and its peripherals form a system but the system is relatively isolated. You can change from one size of floppy disk to a smaller size without ripping holes in the walls of your office. It's the computer that changes, and the things that attach to the computer.

In the world of computers, where individual components are disaggregated from each other, and the computer system is disaggregated from other parts of the infrastructure, change comes very rapidly—as always, the more disaggregation, the more creativity. The answer to the question about why computer standards change is that the standards *do not* actually change—what happens is that people abandon old technologies and adopt new ones. The old standard for 5 ¼ disks is still around; the old standards for various printers, disk drives, and other stuff piled up in my basement haven't been revoked. The world has moved on, and computers don't use these old technologies anymore, the same way (and for the same reason) that a new car doesn't use wheels that are the same as those of a Model T.

The Benefits of Standards

I have listed some of the problems that were solved by introducing standards, but because standardization is a form of disaggregation we expect to enjoy the universal benefits of disaggregation. Let's look to the nut and bolt industry and see whether the universal benefits are there.

A few benefits are quite obvious. *Cost reduction* for the nut and bolt industry came to the customers in the form of less-expensive goods and cheaper repairs. *Competition* increased, with a focus on better and more inventive nuts and bolts rather than the shape of the threads; ordinary nuts and bolts were eventually commoditized, a typical result. *Simplicity*—well, certainly it was a lot simpler for everyone to deal with standard sizes than to deal with the chaos of no standards at all.

However, my favorite benefit is *creativity*, and there would seem to be a problem. If nuts and bolts are standardized, how could anyone innovate? After all, the shape of the threads and their sizes are the same everywhere, so how can you be creative when you manufacture nuts and bolts? Actually, standardized nuts and bolts means *more* creativity, not less. That's because once nuts and bolts were standardized the nut and bolt manufacturers found new ways to differentiate their products—they had to in order to survive the era of commoditization. Some of that differentiation came from the sales and marketing teams, whose job it is to explain superior manufacturing quality, better materials, lower prices, and faster delivery.

The technical staff also contributed. The standard specifies the shape and size of nuts and bolts, but it very carefully leaves lots of room for innovation. Today manufacturers make nuts and bolts of special materials (you can get bolts in steel, brass, nylon, and other odd materials), with special coatings and colors, with novel shapes for the heads of the bolts, and other innovations. Product designers who need nuts and bolts can order all sorts of wondrous shapes and sizes straight out of a catalog, pay far lower prices than if they had to design and build their own from scratch, and can spend their time thinking of creative ways to use these nuts and bolts. Standardization, paradoxically, leads to more creativity.

The nut and bolt industry may have been the first industry to realize how standards would grow their business, but it certainly hasn't been the last. Like all good avalanches, this one kept going and going, gathering more momentum. The idea of standards to lower costs and increase business spread across every industry. The result is today's society with its standardized, interchangeable, and, above all, *repairable* parts.

Chapter Nine

Coping with Surprises

I enjoy a surprise as much as the next person, but I have to admit that there are some surprises I can really do without, such as jury duty; dental work; late night adventures in plumbing repair; or an early morning round of hand-to-mouse combat in the kitchen. I suspect that most everyone will agree that the sudden realization that your company is in the path of an avalanche is something else that isn't very popular. As I hope I've made clear, sweeping technological changes can't be avoided—and because they will happen, you'd best be ready for them. In this chapter, I discuss strategies to cope with the inevitable reality that one day, despite your best intentions, you will be surprised by a technological revolution.

Digital Photography

The primary example I use in this chapter is digital photography. Digital photography accomplishes a series of remarkable disaggregations.

- *Digital photography disaggregates cameras from film.* Before the invention of digital photography, all cameras had to find some way to accommodate film—in a real sense they were built around film, which put limitations on the size and shape of any camera. Digital cameras operate independently of film. Cameras are now electronic components that can be built into almost anything, like cell phones and personal digital assistants (PDAs). (This idea is arguable, by the way; digital cameras must still record the image onto something, and, whereas before they recorded onto film, now they record into computer memory. Is that disaggregation or just plain substitution? But let's grant me this idea anyway. Computer memory is a reusable electronic component that takes up relatively little room; this is very different than film, which is a bulky single-use chemical medium.)
- *Digital photography disaggregates images from the storage medium.* Images were once bound to a medium. They came in tangible formats—negatives, paper, books—and the image was stuck to the medium. Digital photographic images are stored in digital format and can be transferred or copied to another storage location. A more radical statement of this same idea: images were once real, always part of some storage medium recognizable to the human eye as an image. Digital photographs are virtual.

Here is a list of the innovations that went into popularizing digital photography:

- High-quality sensors to convert light into computer signals
- High-speed computer memory to store the images while inside the camera
- Cheap disk drives to store the images when they were downloaded from the camera
- Mathematical algorithms to shrink the size of image files to something manageable, along with microprocessors small enough to fit inside cameras but smart enough run the algorithms

Digital photography is a classic case of synergy, a revolution built out of parts of other revolutionary technology. The revolution didn't happen overnight; anyone who was paying attention had time to prepare for the consequences. And as the cost and size of each item on the link continues to shrink, the revolution is bound to continue.

Five Steps to Remember: A, E, I, O, U

Now we proceed to the discussion. You're heard the rumble of an approaching avalanche—what do you do? What positive actions will you perform to cope? I recommend "AEIOU": *Anticipate, Exchange, Inform, Observe, Undermine.*

A: Anticipate

My family and I don't venture out for a hike in the mountains until we've done a pretty thorough job of checking local conditions. Is it going to rain or snow? What temperatures can we expect? Are there bears nearby? (Bears are cute only at a distance, and grizzly bears are cute only at a *great* distance.) And once we've done all that, we also pack water, spare food and clothing, matches, whistles, flashlights, and emergency shelters. All this for a two-hour hike.

Does this sound a bit extreme? Maybe even laughable? I suggest you think again. I met someone who nearly died in a blizzard, lost and disoriented, who was saved only because he had carried his whistle in his pocket. A friend of mine was out on a three-hour hike with friends. That hike developed into an overnight rescue mission that ended with her being airlifted out by helicopter, and her stash of emergency supplies made all the difference in the world. Thankfully, I've never had anything of the sort happen to me, but I accept the fact that even a short hike in the mountains requires that I carry a significant amount of gear.

Unless they have personal experience or they are willing to learn from others' predicaments, people tend to laugh off worries about improbable emergencies and don't prepare for them. Most companies don't really expect revolutionary change and the obsolescence that follows. The companies that prepare for obsolescence generally only prepare in a limited sense. The major appliance industry—the industry that makes stoves, refrigerators, and washing machines—is certainly prepared for limited obsolescence; after all, the manufacturers update their models just about every year. But that is *model* obsolescence, not *industry* obsolescence. As far as they're concerned, people will always need washing machines to clean their clothes. But what would happen if someone invented a radical new technology that made water-based washing machines obsolete? That would be industry obsolescence, and it's hard to believe that the washing machine industry is really ready for something so drastic.

If you're looking for an industry to emulate, an industry that really knows how to anticipate radical changes—an industry that knows what it's like to have its products become not only obsolete but forgotten entirely—look to the electronics industry. There's always another computer or electronics invention around the corner. Videotape recorders, for example, have entered a downward spiral; they're being replaced by hard drives and DVD recorders. Or perhaps something will come along and make floppy disks obsolete and they'll be replaced by . . . whoops! Too late! Not only do most computers omit floppy disks in favor of flash drives and CDs, but the CD itself is headed for obsolescence over the next few years. Electronics companies are very aware that their products can be rendered completely obsolete in a relatively short period of time, and their businesses are structured to cope with this fact of life. And if they can do it, you can too.

E: Exchange

The core idea that makes a technology revolution survivable for your business is to realize what advantages you have and to capitalize on them. You often have some very valuable knowledge that fits into the new, postavalanche scheme of things, and if you play your cards right you can still prosper. The trick is to distinguish between the skills and expertise that are intrinsically part of the old, preavalanche technology and those that can be transferred to the new technology—and then to exchange the obsolete technology for the new technology. It's a matter of carefully assessing your current knowledge, expertise, and technology and disaggregating the obsolete parts from the parts that are still important and viable.

Automobile engines offer an example of a revolution in progress. There's slow movement toward replacing the standard automobile engine with a hybrid electric system. But that change won't put the current automobile manufacturers out of business if they have any sense at all. They'll swap out old technology for new technology; their knowledge of how to design, manufacture, sell, and maintain automobiles will remain valuable.

In the field of photography, some camera companies survived the avalanche. Camera companies knew how to manufacture and market cameras, and when the avalanche started, they started building digital cameras but didn't drop their film cameras. As sensors improved and prices dropped, the camera companies introduced more and more digital cameras into their product lines; today they offer both film and digital products, and you can choose whichever you please. When film finally becomes completely obsolete, they'll be ready. They are exchanging technologies, many survived the avalanche, and they didn't see their market taken away from them by, for instance, startup companies or computer manufacturers.

The camera companies had some luck. Although they had to know everything about film, their core business is taking pictures; the digital photography revolution affected them, but their core expertise wasn't wiped out by the digital photography avalanche. What about companies that have a core product that's made obsolete—can they survive? Kodak's core business was to manufacture and develop photographic film, and they were squarely in the path of the avalanche with no place to run. But Kodak did have interesting things going for it when the revolution hit:

- Color. Kodak excelled in the chemistry of color, and even had expertise in how color worked on computer monitors.
- Paper. Kodak's photographic paper for making photographic prints was technically excellent and had tremendous brand recognition.
- Connections. All across the world, Kodak's extensive network of partners in the film finishing business faced the same problems that Kodak faced.
- Consumer information. After a hundred years of gathering marketing information, Kodak had a slew of data on the habits of its customers.

The digital photography revolution didn't make film obsolete overnight—use of film declined gradually. Like the camera manufacturers, Kodak had plenty of time to adapt. Kodak could have grabbed opportunities to capitalize on its advantages to avoid damage from the digital photography avalanche. Here are a few scenarios.

- Color. Computer printers have a tough time getting the color of photographs right. Kodak could have started its own ink business or gone into partnership with a printer company to make inks and software that produced accurate colors.

- Paper. Inkjet printers use special paper to print photographs, and Kodak could have supplied that paper as an extension of its photographic paper business.
- Connections. Kodak's partners in the film sales and film developing businesses needed to find some way to pick up the slack when sales started to fall off. Kodak could have provided specialized equipment to their partners—just as they had provided for developing film— to print out digital photographs for customers. In this way, end users wouldn't have to purchase special printers, ink, or paper to get high-quality prints of their digital photographs.

What's interesting is that Kodak did take many of these steps, and did show foresight. For example, in 1998, they opened a partnership with AOL to print film and put digitized copies of the photos online, and did so in the era when digital photography was still just taking hold.

Despite these steps, Kodak kept their film-centric view of photography. For example, Kodak offered Picture CDs and Photo CDs; the idea was that people would buy film but Kodak would digitize the photos for them. These products let people have digitized photos but still kept them away from digital cameras—Kodak apparently wanted to prevent customers from abandoning film entirely. When I look through Kodak's year-by-year list of corporate milestones from the time of the invention of the digital camera onward, I see some important advances toward the digital age, but I also see a continued strong emphasis on film.

Kodak finally shook off its attachment to film in late 2003. They stopped investing in film, decided to sell off some of their obsolete businesses (e.g., the Carousel slide projector business), and concentrated on the digital photography market. Since then, they seem to have made decent progress against a slew of well-entrenched competitors. On a trip to the office supply store and the electronics store down the street, I can find Kodak computer paper for printing photographs at home, Kodak brand printers, and Kodak ink. Your local drugstore or photo supply shop might have a Kodak-built system for making prints of your digital photos. Kodak sells digital cameras that interface easily with Kodak's specialized photograph printers.

As of today, Kodak has successfully exchanged its old technology for new technology. They leveraged their expertise, connections, know-how, and especially brand recognition; they developed new ideas. They no longer dominate their field, but at least they're holding their own.

Polaroid didn't make it. Polaroid had advantages similar to Kodak's; I could repeat the entire list and it would look much the same. Polaroid had a wonderful market niche, instant photography. Polaroid's name was everywhere, they made a nice profit from giving away cameras and selling film, and every business I've ever worked at had a Polaroid camera somewhere around the office. But instant film has essentially vanished, replaced by pictures available instantly through digital cameras. In theory, Polaroid was perfectly poised to move from supremacy in instant film photography to supremacy in instant digital photography; after all, the end product was the same—an instant photograph—and they could have made the transition from selling film cameras with self-developing film to selling digital cameras and printers. In practice, Polaroid declared bankruptcy in 2002. Polaroid still has a market presence and has Polaroid-brand digital products, but the company is a shadow of its former self.

I: Inform

When Kodak announced its plans to sideline the film business and compete in the consumer and high-end print business, many people were skeptical that they'd be able to resurrect the company—after all, they were entering a field with lots of competitors. Some shareholders were absolutely livid; according to them, Kodak was a film company and by golly would always remain a film company. They issued scathing comments that rejected the very idea of competing in the digital photography market—they thought that Kodak's plan would be a complete waste of money, and Kodak might as well "take the free [available] cash and set it on fire" instead.

These shareholders failed to block Kodak's plans and Kodak proceeded with its so-far successful reform; in the end Kodak got off lightly in this little revolt. Sometimes the problem of resistance to change can be much worse. I've had my own run-ins with this "let's not change anything" attitude at more than one company; they're never fun and they're never helpful.

A little resistance to reality can go a long way toward sinking a company. What's to be done? You must inform, educate, and persuade. Your audience isn't just senior management; your audience is your sales team, your engineering team, and the ultimate owners of the business—your shareholders, if you have them. Don't forget your business partners, your bank and other equity holders, industry analysts, and the press.

O: Observe

Kodak had an interesting advantage as it entered the digital photography business: it could benefit from everyone else's mistakes. Kodak did stay in the game during the entire revolution with its own forays into digital photography—the "Picture CD" and online photo Web site spring to mind. By the time of their 2003 decision to virtually abandon film and focus on digital, many other companies were already in the market and earlier mistaken ideas had already been discarded.

Kodak did a very credible job of learning from other companies' experiences. The home printing market, as of the time of this writing, has settled into a wacky but typical business model: light-duty printers are free or close to it, but the ink cartridges are very expensive (a wine bottle filled with printer cartridge ink would cost $900). When Kodak entered the market, they had two choices: they could accept the current industry model or attempt to introduce a new one. They wisely decided to compete on their strengths. Kodak sells "photo printers" that do nothing but print high-quality, durable photographs; to make them easy to use, you can attach your Kodak camera directly to the printer to transfer the photos. They sell Kodak ink and Kodak paper for these printers and charge a pretty penny for them. At the same time, they're nibbling at other parts of the market by selling photograph paper for use in other companies' inkjet printers. The end result is an impressive performance by the company, which (at least at present) has carved out a nice profitable niche for itself.

There's an old adage that goes, "Experience is what lets you recognize your mistakes when you make them again." If your competitors paid heavily for experience, there's no reason for you to make the same mistakes as they did. You will have to work hard to get the information, experience, knowledge, and wisdom that they accumulated. Learning from someone else's mistakes is certainly not free—but it certainly is invaluable.

U: Undermine

Only under pressure, more or less at the last minute, did Kodak decide to sharply reduce investment in its old core business of film and concentrate on building products for the new digital era. What took so long? It's really hard to know; it seems to have been a combination of inertia and nostalgia, at least from where I sit. Nostalgia is fine and good, and I understand their feelings; but a large company that's structured for steady or increasing sales cannot retain a declining business. To survive, Kodak had to abandon almost all further development of their old film business after stripping away any knowledge, technology, or experience that they'd need for new technology. Kodak's new business undermines their old businesses—they're trying hard to make it easy to move away from film and into digital photography. Kodak made a dramatic turnaround, but then again if they had undermined their old business earlier they wouldn't have *needed* a dramatic turnaround! You must be willing to undermine your old business to focus on the new ones.

They way I usually phrase this advice is, "Make your own products obsolete, or your competitors will do it for you." When the avalanche hits and your old business is in decline, you have to decide what you want to do about it. You could transform your company to an end-of-life business; this is a very legitimate choice, but note that I use the word *transform.* A company that sells a product in slow decline needs a structure different from that of a company with a steady or increasing market. Another option is to look for an entirely new market for your old product; that trick sometimes works if you're clever enough or desperate enough. Or you can decide to undermine your old business: forget about your old technology and adapt your experience, your business relationships, and your other technology to the new reality. What you can't do is ignore the problem and hope it goes away—because it won't be the problem that goes away.

A Little Pep Talk

Gutenberg didn't invent the printing press, but he did introduce movable type to Europe, which was the start of a fantastic revolution in printing technology. The innovation was a disaggregation: movable type broke the connection between the printing plates and particular pages being printed. Movable type let printers quickly and easily set type for each page as needed, instead of literally carving out an expensive plate that could only print one particular page.

But here's a cheerful thought: Gutenberg's invention affected just one part of the printing press, the plate. There were printing presses before Gutenberg and there were presses afterward. And both before and after, there were paper suppliers, distributors, writers, markets, salespeople, and business partners. Gutenberg transformed one crucial part of the technology and triggered a revolution, and everyone had to adapt to his invention or be priced out of the market. But the rest of the business, the rest of the special expertise and knowledge about printing, was vitally necessary as product exploded. Craftsmen still built, operated, and repaired the presses; salespeople dealt with customers; purchasing agents bargained with the suppliers for paper and ink; and bookbinders still sewed the pages and bound the covers. Gutenberg's revolutionary innovation didn't kill off the old printing business—although they had to exchange some old technology for new, the rest of their expertise became even more valuable.

When I first began to research this book, I hoped to reveal some magic formula in this chapter, some sure-fire technique to deal with revolutionary change. What I've found instead is very straightforward: use disaggregation to cope with disaggregation. Turn your sight inward and generate innovations. What parts of the business are truly obsolete? What techniques, knowledge, ideas, attitudes, relationships, and practices can be disaggregated from the parts affected by the revolution? If you can solve that problem, you can cope with the revolution, just as Kodak, the printers of the Middle Ages, and countless other businesses have done before you.

Chapter Ten

Marx, Lenin, and Gates: Failed Counterrevolutions

Stopping an avalanche is a losing proposition, but that hasn't stopped people, companies, and nations from trying. In this chapter, I'm going to discuss the strategy of resistance—the strategy of trying to stop or reverse disaggregation. There have been a few successes. Well . . . not really successes. I don't think I've found any examples of permanent victories over disaggregation. What I've found is that suppressing disaggregation usually relies on force, and, although force is very effective in the short run, in the long run it's almost impossible to permanently suppress a good idea. Let's look at some of the examples, starting with an experiment in reaggregation that killed millions and millions of people.

Karl Marx, Vladimir Lenin, and the Nuts and Bolts of Socialism

One of the worst regimes of the twentieth century was that of the Soviet Union. The Soviet Union dominated an empire that stretched across Asia into Europe. The government deliberately murdered tens of millions of its own citizens and sent many others into a vast system of internal prison camps. Why? What drove the Soviet Union to such extreme behavior?

They weren't simply sadists. The problem was that Soviet ideology, Marxism-Leninism, required that the Soviet Union overturn two fundamental disaggregations that are crucial to modern civilization: corporations and money. To reverse these disaggregations required massive force because doing away with the disaggregation meant doing away with the benefits. Just what are the benefits of corporations and of money?

Corporations, Trust, and Sharing

A modern semiconductor factory costs several billion dollars. How can any one person possibly afford to build or buy a semiconductor factory? The answer is that no one person can and no one person does. People form groups to raise the money, and the fundamental structure people use to manage and spend this kind of money is the corporation.

Corporations disaggregate both ownership and authority to solve the problems of large-scale enterprises. Disaggregation of ownership solves the problem of how to define ownership when there are millions of owners. Corporate rules govern how each shareholder exercises their property rights—for example, at the end of the year when it's time to divide the profits, shareholders can't just walk into the factory and make off with whatever they think is their fair portion. The disaggregation of ownership leads to sharing, and the corporation lets people who will never even meet each other pool their money, create enterprises, and generate more wealth.

Disaggregation of authority prevents chaos. When I buy stock in a corporation, I become a partial owner and I'm entitled to certain rights—but I can't waltz into the factory and tell the workers what to do. Corporate rules about who has what authority, despite their ownership in the corporation, leads to trust that the corporation will be run properly.

The creation of the corporation resulted in thousands of innovative ideas in the world of finance—new ideas on how to share ownership and build trust. If you want to build, equip, and staff a factory, you have dozens if not hundreds of financial options to choose from.

Karl Marx proposed to undo these disaggregations and Vladimir Lenin carried out the plan. Marxism-Leninism had no room for corporations; workers had to own the factories, or rather the state would, acting as proxy. Both ownership and authority were reaggregated: the State built, owned, and operated the factory. By disposing with corporations and all the creativity they allowed, Marxism-Leninism made the financial problems of building decent factories almost insoluble. Disaggregation of authority creates trust; it follows that reaggregation of authority dissolves trust, and the disastrous record of state-run factories completely justified that lack of trust.

Even more damaging was the Marx/Lenin doctrine that tried to undo money.

Money and Information

You know that phrase "reinventing the wheel"? Some innovations are so basic to human society and are so important that they've been invented and reinvented time and again throughout history. The wheel would seem to be one of them, but it isn't—human cultures that were cut off from Europe and Asia, such as the civilizations in the Americas, never invented the wheel. What's more important than the wheel? *Money.* Money has been reinvented, time and again, all across history.

Money disaggregates value from the thing itself, and this disaggregation has been applied to all sorts of things—from goods, services, and intangibles (e.g., fishing rights) to concepts and emotions ("pain and suffering"). How money works is pretty straightforward. People pick some scarce resource—salt; beads; cowrie shells; large rocks; brightly colored feathers; or precious metals, such as copper, silver, and gold—and trade those scarce items instead of the actual commodities. All that matters is that the "money" be scarce; what's scarce is irrelevant. The United States dollar, for example, isn't backed by silver or gold; objectively speaking, about 10% of the U.S. money supply consists of worthless pieces of scrap paper and the remaining 90% is even more worthless electronic bank records. What makes the U.S.

dollar work is the U.S. government. The government keeps the supply of dollars sufficiently scarce, everyone accepts the fiction that U.S. dollars are valuable—and thus they are, because people value them.

International commerce, banking, and the stock market all would be impossible without money. Money makes everyday life easier as well; it's relatively simple to design a vending machine that accepts coins, but a vending machine that accepts live chickens would be much more difficult.

A lesser-known but absolutely vital function of money is to provide prices and thereby to quantify value. Prices are an interface, in our sense of the word—the economy consists of thousands of disaggregated businesses, and we know that disaggregated pieces require an interface to let them work together. Prices provide that interface; prices provide a way to convey extremely valuable information between businesses.

Let's say I own a tractor factory. Since I need nuts and bolts for the tractors, I call a few nut and bolt manufacturers and outline what I need: the sizes, schedule, and quality. The manufacturers will offer bids. If my requirements are too strict or if I've asked for something difficult or too unusual, the price will be higher and that will tell me that I've either got to adjust my requirements or give up some of my profits. But my nut and bolt problems are solved—I specialize in building and selling tractors, and someone else specializes in making and selling nuts and bolts. Money and prices provide the interface and let us work together.

What's really amazing is that the process of supplying nuts and bolts for all of society's needs is managed automatically through the price interface. Except in extraordinary circumstances, who has ever heard of a nut and bolt shortage in a modern capitalist society? If the demand for nuts and bolts goes up, the factory owners expand or build new factories. If the demand goes down, factories lower their prices, find new markets, or close. If a factory produces lousy products, fails to meet its delivery schedules, or provides poor service, customers leave and the problem corrects itself one way or another.

Marxism-Leninism and the Lack of Prices

The Soviet Union's Marxist-Leninist government transformed prices into a meaningless exercise, a number set by governmental fiat. It's not that they forced everyone to dispense with money; what they did was destroy the relationship between value, as measured by prices, and the underlying objects. With price's information gone, the interface to allocate goods and services was gone. The Soviet government substituted central economic planning for prices—which failed, miserably. The Soviet Union had vast reaches of fertile farmland but couldn't feed their own population. They had some of the greatest deposits of natural resources in the entire world—timber, oil, gold—but were incapable of exploiting them. Ask someone who lived there in the "good old days" of Communism about the shortages and shoddy quality of even the most basic consumer goods.

Let's say I live in the Soviet Union after the Marxist-Leninist revolution. How would I manage that tractor factory? When I need nuts and bolts, I request them from the Glorious Heroes of the Revolution Nuts and Bolts Factory, which produces them according to quotas set by the State Planning Committee under the authority of the Council of Ministers. This system leads to two immediate problems. The first is that, with quotas set by government agencies, I can only hope that the agency anticipated all of society's needs in advance when it laid out the current Five-Year Plan.

Any error could mean widespread disruption in the economy. It goes without saying that, even in an honest and open society, perfect information and perfect forecasting would be impossible; in a police state like the former Soviet Union, the idea would be laughable if the results weren't so utterly tragic.

The other immediate problem is that the nut and bolt factory doesn't earn a profit by supplying me with wonderful products and excellent customer service; their goal is to meet—or, better yet, evade—their sometimes wildly unrealistic quotas with the absolute minimum effort. (Under Stalin, another important job goal of managers was to avoid being shot, as an example to other managers, for poor performance.)

How did this play out in real Soviet life? The shoe industry provides a hint. The shoe manufacturers turned out plenty of shoes each year, three pairs for every person—but people still waited in line to buy shoes, mostly imported ones. The problem was that Soviet shoes were the wrong size or wrong style. The factories found it easier and more economical to set up their machines for a single size and style and run off as many pairs as they could get away with. If the quota was set by the number of shoes, smaller shoes were easier to produce; if the quota was set by tonnage, fewer but larger shoes were easier to produce. And as for what people actually wanted to wear in terms of style and color, that wasn't relevant to the quota.

Back to my hypothetical tractor factory. The nut and bolt industry also produced according to quotas—which meant that my factory couldn't rely on their shipments. As manager, if I can't rely on nuts and bolts from elsewhere and I have to deliver my quota of tractors, I'll manufacture my own instead. That's exactly what happened, in industry after industry. In the Soviet Union, no more than half of the nuts and bolts were made in the official factories. The rest were made in captive facilities—an industry, such as the tractor industry, would set up its own plant to make nuts and bolts. Hard as it is to believe, one of the earliest disaggregations of the modern era, that of nuts and bolts, became undone.

Every industry was cut off from every other industry, as though they were stranded alone on a desert island, and, like castaways, each industry built the necessities of life from scratch. The simple commodities that Western manufacturers take for granted—nuts and bolts, bricks, paint—were made in endlessly duplicated facilities. Aside from the horrible inefficiency, I believe that this lack of disaggregation had another effect. No disaggregation means no benefits of disaggregation, and in particular the benefit of specialization—in this case, manufacturing expertise in how to make nuts and bolts, bricks, and paint to the highest quality standards. Back in my graduate student days, as we would build experimental equipment in the machine shop, we used to joke about the poor craftsmanship of Soviet products: the hammer and sickle on the Soviet Union's flag stood for their low-tech production technique, "cut to fit and beat to match."

We were wrong about the hammer and sickle. Marxism-Leninism's inefficient system of central planning brought death, starvation, terror, deprivation, and poverty to the Soviet Union. Creativity, synergy, competitiveness—all of these were inevitably damaged or destroyed with the imposition of Marxist-Leninist doctrine. The unnatural act of rolling back basic innovations and their benefits led the Soviet government to oppress their own people because such a colossal failure could be imposed only by a police state. The hammer and sickle ultimately came to mean, "Hammer your population and cut down the ones who resist."

Bill Gates: Running the Wheels Backwards

I've seen Microsoft in action, and it's really, really not very pretty. A few years ago I was at a conference for speech technologies. We were excited over a new standard for speech technology from the World Wide Web Consortium (W3C), one that would make it infinitely easier to write applications; easier applications would mean more applications, and that would be mean more business. To my great dismay, Microsoft came along with their own "standard" and muddied the water. Their marketing material—handed out during the presentations!—deliberately attacked the W3C standard, a breach of professional decorum that absolutely shocked me. The materials focused on fear, uncertainty, and doubt: which standard should I adopt; is it true the W3C standard is incomplete and unsound; which one will be adopted by customers, the one created by the speech industry, or Microsoft's "standard"? A colleague who'd worked hard on the W3C industry standard was almost speechless with anger.

When I talk to people who work for Microsoft, they tell me that they make the best products in the world. In contrast, government prosecutors and private companies all over the world have taken Microsoft to court over their business practices and have complained about illegal behavior. If their stuff is so good, why does Microsoft act like a bully? I've heard the usual theory, that Microsoft has an aggressive internal culture that leads them to push the envelope and do things that other companies might avoid. I think it's a more fundamental problem, the same problem that the Soviet Union had. I think Microsoft is attempting to run the wheels backwards and do away with innovations they don't like, and the only way to accomplish that is to use force.

Here's the irony: Microsoft was founded on a series of fundamental disaggregations, and now they're trying to do away with some of them. One of the most fundamental disaggregations of computer science today is the disaggregation of operating systems and applications. Applications are the programs that do the useful things you bought the computer for in the first place: writing documents, browsing the Web, and playing video games. The "operating system" (e.g., Windows XP) takes care of the housekeeping of the computer: it figures out what you're typing at the keyboard and is in charge of keeping track of files on the disk drives.

The separation of operating system from applications offers all the usual benefits we expect from disaggregation. It makes for simplicity because each program can focus on doing just one task correctly; for more reliable programs because an application error shouldn't crash the operating system and thus the entire computer; and for better security because the operating system can defend the computer against rogue applications.

And then there's that problem benefit: competition—at least, it's a problem from Microsoft's perspective. When application programs are separate from the operating system, people are free to choose which programs they want to purchase—just because they purchase a Microsoft operating system doesn't mean they absolutely must purchase Microsoft applications.

If you control the operating system, there are tactics you can use to reaggregate applications with the operating system and to nudge customers toward buying your applications. First, there's outright sabotage: change something in the operating system to make applications from other companies fail. Microsoft was accused of this for years. Another tactic is to change the way the operating system works with applications—the so-called API, or application programming interface. If Microsoft makes a few changes in the API and simultaneously releases new versions of its software, Microsoft's software will work—while other software vendors scramble to fix their software when they finally discover what's happened to the API. Then there's the "hidden API" tactic, where Microsoft keeps some of the API secret. Competitors worry that Microsoft developers can use the hidden API to build speedier or slicker applications.

And the most retrograde tactic of all is barefaced reaggregation. In the most infamous and blatant example, Microsoft integrated the

Internet Explorer Web browser into the Windows operating system. This maneuver landed Microsoft in hot water; after a protracted court battle with the U.S. Department of Justice, Microsoft ended up with a slap on the wrist and the lion's share of the Web browser market.

Embrace, Extend, Extinguish

These tactics show how someone with monopoly or near-monopoly control of a crucial piece of technology can fight off disaggregation, but actually the lesson goes deeper than this. Microsoft gave Internet Explorer away for free even though Microsoft's competitors were, for the most part, charging for their Web browsers. Why did Microsoft use reaggregation to try to lock other browsers out of the desktop? If Microsoft was just giving away the software anyway, why did they care about competing browsers?

Web browsers are a gateway to the World Wide Web and the rest of the Internet. If Microsoft were to gain complete control of the Web browsers, they could use that control as a wedge to capture the World Wide Web—to take it away from the international community and bring it under Microsoft's control. This tactic, which perhaps you might recognize from your industry, is often called "embrace, extend, and extinguish." If you're the dominant player in the industry, here's how you can use this tactic.

- *Embrace* an industry standard by incorporating it into your product.
- *Extend* the standard by adding nonstandard, proprietary extensions under your exclusive control.
- *Extinguish* the standard. Make the proprietary extensions important or vital in some way—for example, make them part of your other products, and use your market power to push them into the marketplace. Eventually, anyone who wants to be part of the marketplace has to adopt the extensions. The industry ends up with a "standard" that is completely controlled by you, the dominant player. Competitors struggle, but lack of access to the proprietary technology in the "standard" places them at a disadvantage.

In the case of Internet Explorer, Microsoft added nonstandard extensions—well, I should be careful. When I talk to folks from Microsoft, they call anything that Microsoft uses an "industry standard";

what I mean by "nonstandard" is that Microsoft added its proprietary extensions. Some businesses followed Microsoft's lead and started to use the proprietary extensions, which had two implications. First, people had to use Internet Explorer to view those Web pages. Second, the Web pages couldn't be viewed at all on computers that used the Unix operating system. I used to run across Web pages like that from time to time, and all I could do was throw my hands up in the air.

That's a crucial point—part of the Web became unusable if you didn't use Microsoft products. The rest of the scenario is easy to project: at some point, Internet Explorer would start to use proprietary technology to interact with Web servers—Microsoft Web servers. If everyone uses Internet Explorer, and Internet Explorer uses proprietary technology that works only with Microsoft Web servers, then all businesses on the Web must use Microsoft's Web server software, which does cost money and generates a nice profit for Microsoft. Fortunately, this never came to pass: the Internet standards remained intact. At the time of this writing, the vast majority of Web servers don't use Microsoft Web server software, Internet Explorer uses standard technology to interact with Web servers, and most businesses don't use the proprietary features of Internet Explorer—because their customers complained.

Another way to use the concept of embrace, extend, and extinguish is to sabotage competitors. Microsoft tried this against the Java programming language, which Microsoft quite correctly viewed as a threat. The Java programming language, from Sun Microsystems, lets you write a program once and have it run, without any changes, on any computer: Windows, Macintosh, or Linux. Java is even built into many cell phones. This undermines the Windows operating system: if the application is completely independent of the operating system, why bother with Windows? Microsoft *embraced* Java and then *extended* it with Microsoft-specific stuff, in violation of their Java licensing agreement. The extensions made Microsoft's Java incompatible with other Java, with the result that programs lost the portability between operating systems that was the whole point of Java in the first place.

Microsoft had picked a fight with Sun Microsystems, the owner of Java and a company with deep pockets of its own that could afford to take Microsoft to court. After years of litigation, Microsoft and Sun eventually settled; Microsoft paid Sun about $2 billion to settle various antitrust and patent claims.

The Open Source Challenge to Microsoft

The open source software movement has a unique answer to Microsoft's counterrevolution: they're fighting reaggregation with more disaggregation.

Microsoft is a classic software company, and classic companies all share certain features: in-house experts ("gurus") decide what the software should do; legions of programmers write the software's "source code"; the source code is a guarded, highly proprietary secret; in-house specialists test the software; marketing finds customers; and technical staff members write documentation and help customers. Managers exercise strict control over productivity and schedule.

The open source movement challenges classical software companies. The open source movement has a completely different philosophy, which in turn leads to a completely different way of writing software. The open source software movement believes that source code should be available for public scrutiny, and many adherents believe, in addition, that software ought to be freely available—literally without charge. These ideas lead directly to a new way to produce software, a kind of disaggregated software company: designing, writing, testing, documenting, marketing, selling, distributing, and maintaining software are all separate functions and can be performed by separate individuals who don't necessarily coordinate their activities. The work isn't necessarily performed by companies, even though it often is; fantastic amounts of work are done by volunteers who are simply interested in the software.

The most famous open source project is Linux, an operating system that can be used instead of Microsoft Windows. Linux dominates the Internet infrastructure. Linux continues to make inroads into the back-office operations of businesses, a market that Microsoft lusts after. Linux even poses a terrific challenge to Microsoft Windows for the ordinary desktop computer because of all the advantages Linux offers.

- *Lower Costs.* Windows is a single-source product available only from Microsoft, which allows Microsoft to control the price of

Windows. Linux, on the other hand, is available from many different places, and many versions of Linux are available for free. These free versions provide the baseline against which any for-profit Linux product or business must compete, and the result is that Linux software is less expensive than Windows software.

- *Better quality.* Linux is different than traditional software projects. Most software companies have relatively few people to review the software or test it. Even if the company has thousands of programmers, the programmers all work on different projects or sections of the software, and any particular piece of code is seen only by a relative handful of programmers. Since the source code is a proprietary secret, only company insiders can debug the source code. In contrast, Linux doesn't hide its source code; ownership is disaggregated from any individual. Every time software is released, thousands of knowledgeable individuals compile and run the latest version. With thousands of "eyeballs" looking for errors, it's not surprising that Linux finds and fixes bugs rapidly; the fixes are transparent—no mysterious, unexplained changes.
- *Superb security.* Linux has major advantages over Windows in the area of security; here are three of them. The first is that Linux is designed in a disaggregated but orderly fashion, so security errors remain isolated, unlike Windows. Second, Linux has thousands of eyeballs—knowledgeable users who test Linux in a huge variety of settings and with many different tools. Security errors tend to be caught quickly and fixed quickly. Third, anyone who is concerned about potential security flaws can inspect Linux software for themselves — they don't have to rely on assurances from others.

If you want to stop a project like Linux, which competes directly with Windows, there's no single company to crush or buy out—thousands of experienced programmers around the world contribute time and energy to Linux, and they can't be bribed or intimidated. How can Microsoft defend its monopoly against this disaggregated structure?

One Microsoft tactic has been a campaign to discredit Linux. The fun started with a list of weird accusations against open source: Open Source software is a "cancer"; Open Source software is un-American; Open Source endangers your company's intellectual property. After that stage fizzled out, Microsoft started on the next round, which purported to show that open source software's quality isn't as high as Microsoft's and that it costs more; critics and governmental authorities denounced Microsoft advertisements as misleading. Lately, Microsoft claimed that their Shared Source program is the same as open source, while casting aspersions on open source. This is a classic FUD campaign—casting fear, uncertainty, and doubt—and Microsoft is very good at it.

Which will win, Microsoft or Linux? Linux continues to gain momentum; at the same time, Microsoft is fiercely defending its turf on the desktop. Rather than fight power with power, the open source movement in general and Linux in particular follow a different path. Open source disaggregates the entire model of how software is written, and has created a production method that can never be suppressed or dominated; they've shifted the battleground. They've left the monopoly frantically trying to stem the avalanche before the avalanche sweeps aside the monopoly's entire business.

Chapter Eleven

The Role of Government

"All bad precedents began as justifiable measures."

—*Julius Caesar*

If you're worried about competitors with new ideas, I highly recommend legislation. Legislation is an absolutely terrific method to quash upstart innovators. State power applied against your competitor's ideas can prevent disaggregation from taking hold, or reverse a trend in your competition's favor. The automobile industry and the liquor industry have made both excellent use of legal maneuvers to forestall disaggregation.

The threats to the liquor industry and the automobile industry came from the avalanche of change unleashed by the World Wide Web. I live in Chicago, and one day I went looking for a bottle of Old Peculier beer. In the days before the Web, I didn't have many options;

I could call local liquor stores or look at ads in the local newspapers. Getting a price from a liquor store in New York or Montana was just about impossible. The Web changes all that—the Web disaggregates *location* from *information*. A liquor store in Wyoming is only a mouse click away.

The Web provides another benefit: innovators can use the Web to disaggregate commodities from their supply chain. The supply chain of manufacturer, distributor, and retailer can be replaced by a simple, direct connection between any level of the supply chain and the purchaser.

The liquor industry is vulnerable to disaggregation of its supply chain because wines are intrinsically commodities. A bottle of wine is a bottle of wine, no matter where you buy it. That's not strictly true, because wine stores vary greatly in the selections they offer, the experience level of their staff members, and how well they treat the wine—but only oenophiles like me care very much. Many customers are just as happy to save some cash, so why shouldn't they purchase the wine online?

The opportunity to break up the liquor business supply chain is especially important and very enticing. A silly set of laws, rules, and regulations restrict liquor sales all across the United States. Let's say you want a bottle of wine from a particular vineyard but your local liquor store doesn't have it available. If what you wanted was something else, say some cheese to go with the wine, the store could order whatever you want. But the wine might be just plain impossible to get; in many states, the merchant must order all their wine through a state-accredited wholesaler, and if the wholesalers in your state don't carry that wine, you're simply out of luck. I read a newspaper article about beer sales in Pennsylvania that I still find hard to believe: single cans, up to a limit of one dozen cans, come from one kind of store; more cans or entire cases are available only from another kind of store. You have to wonder whether the legislators spent too much time sampling the product before they came up with that one! Web-based sales disaggregate this entire Byzantine supply chain.

The established members of the liquor supply chain fought back with legislation against Web-based sales—this was particularly easy in those states where all the liquor stores are state-owned, because those legislatures were happy to protect the state monopolies. Some states outlaw shipments from out-of-state; others mandate that all sales take place face-to-face. Some states entered into "reciprocity" agreements to allow shipments between member states. The rather thin excuse underlying many of these restrictions is that Web-based sales would lead to underage drinking. The restrictions are under constant challenge, mostly by out-of-state wineries that are shut out of the supply chain and can't distribute their wares. After the U.S. Supreme Court ruled that out-of-state wineries must be treated the same as in-state wineries, at least one state outlawed all direct-to-consumer shipments. Despite the Supreme Court's ruling, the battle over the supply chain is far from over.

The automobile industry is also ripe for disaggregation of its supply chain. After all, a car is a car; it's from the same manufacturer no matter where you buy it. Why not shop for cars online in the same way we shop online for anything else? Automobile manufacturers, or perhaps a few large nationwide retailers, could let you shop online and deliver the car to your door. Customers could compare prices, service policies, and perhaps the reputation of the warranty repair depots. According to the U.S. Federal Trade Commission, direct automobile sales would cut about 10% off the price of a new car.

The automobile dealers decided to fight Web-based sales through legislation. In response to pressure from the automobile dealers, just about every state in the United States amended their franchise laws to restrict automobile sales over the Internet. The automobile dealers did have to overcome a little problem: they had to explain why the laws served the public good—bills with such titles as "A Law to Enrich Automobile Dealers" tend to be struck down by the courts. For the franchise laws, the ostensible reasons were that if sales were made over the Internet you'd have no place to get your new car repaired, and, furthermore, national dealerships would infringe the dealers' franchise agreements. The laws have several variations, but they share a common result: you can't just log in, pick out a new car, and have it delivered directly from a national dealer. Sales are made through local dealers. At a minimum, this ensures that local car dealers still get the revenue from new car sales; depending on local laws, the Web might not be able to influence prices at all.

Here's an example. I just went on the Web to pick out a car in the Chicago area, something nice to drive around in. Most Web sites offer to send me quotes from Chicago-area dealers if I give them full contact information, but they won't actually tell me what the price is. Even the ones that give a price won't let me just buy the car—a dealer has to contact me. One site did actually offer to sell me a new car, but they wanted me to drive to Wisconsin to get it. Clearly, new car dealerships in Illinois don't worry very much about pesky online competition.

Other industries have tried to take the same route as the automobile industry. When an industry can't come up with a plausible excuse, the court system will sometimes overturn restrictive legislation, as happened to the funeral industry. Federal rules require that funeral homes disaggregate their services and list prices for each one separately. A few customers realized they could purchase coffins over the Web and thereby lower funeral costs; the funeral industry petitioned state legislatures, which passed laws to require that all coffin sales be made by licensed morticians. Apparently the industry couldn't muster an argument they could repeat with a straight face in court—sellers and buyers challenged the laws, and today only a handful of states still restrict casket sales.

The Other Government

Legislatures are very powerful, but there's another branch of government that has tremendous impact on day-to-day business operations. Regulatory agencies can propose rules and regulations, pass them into law, and enforce them, and even have courts to punish anyone accused of wrongdoing. Thousands of regulatory agencies exist at the state, local, and national levels of government. Regulatory agencies can make or break a business by controlling the conditions under which it operates. What can a business charge for services? How must it cooperate with other members of its industry? What are the requirements to enter the industry? What paperwork does the business have to submit? What technologies are permissible? Regulatory agencies are supposed to be independent, nonpartisan guardians of the public interest, but, like the legislatures, they can be swayed or lobbied by the incumbents of the industry.

For a good example of how regulatory agencies deal with disaggregation, competition, and the like, look no further than the Federal Communications Commission (FCC) and the local state regulators, the public utility commissions. Here's a short list of some of the questions before the FCC and the public utility commissions, all of them about disaggregation. The FCC may propose more disaggregation in the Internet service provider industry: if you have cable TV, should the cable TV company be the only company allowed to provide Internet service over that cable, or should they be required to let competitors offer the Internet service? What about your local telephone company? Should it be allowed to force you to pay for a telephone connection when all you want is DSL—why aren't the two services disaggregated? Let's say a cable company provides you with Internet access and also offers Internet telephony; however, you chose a different Internet telephony provider. Should the cable company be allowed to block the data packets that go to your Internet telephony provider and therefore force you to use only the cable company's Internet telephony service, or should the Internet service and the Internet telephony service be disaggregated?

The FCC is heavily involved in the telecommunications revolution and in the Internet revolution; by necessity, it finds itself in the thick of battles over disaggregation. Companies continue to battle over these issues with each other and the FCC; your business newspaper will provide you with almost daily updates on the progress of the telecommunications revolution.

If You Lose Anyway

This isn't the place for a long, detailed explanation of how to lobby legislatures and regulatory agencies; other authors provide excellent books on the topic. I will give this warning: entrepreneurs cannot ignore government. In 1995, during the height of the good old days when high-tech companies were innocent and the government didn't understand the Internet, Microsoft spent $16,000 on political contributions. In 2000, it spent $4.7 million, and in 2004 it spent an estimated $18 million in both contributions and lobbying expenses. That's quite a bit of change, and Microsoft is definitely expecting return on the money. Like it or not, the political system runs on cash, just as it has for all of recorded history. You may need to include lobbying expenses and political contributions in your business plan.

I'm sorry to say that "the good guy"—that's you, the person with the innovation—doesn't always win. Sometimes a monopoly squeezes you out of the market, as we saw in the previous chapter. Sometimes an agency regulation makes your business model untenable, or the legislature just outlaws your business ideas outright.

Historically, people haven't had much of a choice but to accept defeat. In the year 1561, a coppersmith in Nuremberg invented an improved metal lathe. The local guild told him he couldn't sell the lathe, make any more of them, or even leave town without permission. A few years later a local artisan made a copy of the lathe—and was promptly thrown into jail "to teach him not to do it again." In that era, a twenty-mile trip to the next town was a wild adventure, and once your local town banned your invention you were pretty much out of luck.

Today it's different; today we live on a big planet. Ideas, people, and capital flow from country to country, and take root where they can flourish. Each country has a different view of technology, rules, regulations, and how society works. You may be shut out of one market, but you're never shut out of all markets. A good idea is very hard to kill.

Chapter Twelve

Predictions: Three Revolutions in Progress

Question: I've got names and addresses stored in my PDA, in my cell phone, on two different computers, and on at least two different Web sites; none of them are synchronized with each other. Why not?

Answer: At least a half-dozen different industries are battling to control your address book.

Question: Why does the phone ring when I'm watching football games?

Answer: Your phone isn't smart enough to realize you're watching the game. That's about to change.

Question: How does *The Wall Street Journal* manage to publish a print edition, a Web edition, and a mobile edition? Because each edition is different, don't they have to rewrite the story three times?

Answer: At first they had to rewrite the stories, but today it's a lot easier—now that they have started to use some pretty significant technology in their back offices.

Each of these questions and answers comes from a revolution in progress. One started when a quiet innovation triggered multiple avalanches and at the same time founded a new industry. Another avalanche is rapidly sweeping competitors from the marketplace but will likely cause its current round of investors to lose billions of dollars. One revolution is only just beginning—and might never get off the ground.

XML: A Universal Language for Data

The Wall Street Journal publishes a print edition, a Web edition, and a mobile edition (for cell phones and PDAs). They also have an instant message service, an RSS news feed, and will no doubt add new technologies when they become available. All of these editions are different from each other. This graphic illustrates the preavalanche method of publishing these stories:

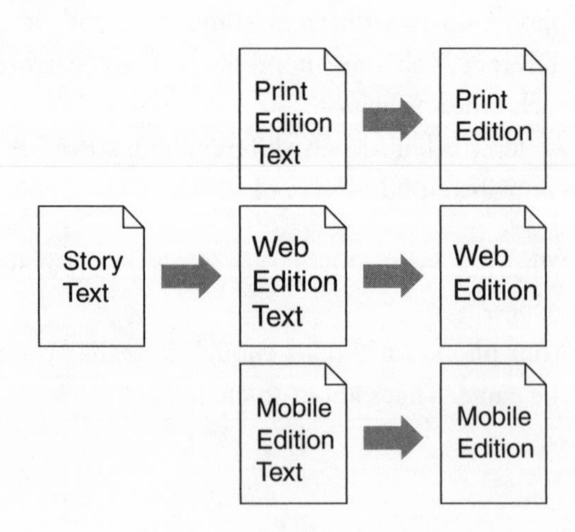

First a writer would create the story. Because of the differences between editions, editors would take the story and create a separate version for each edition, and then they would format the text separately for each edition. Although some of this process was automated, it was still a big bother. And, of course, quite expensive.

The *Journal* realized that there are really two tasks here: one task is to create "information," and the other task is to create the proper "presentation." "Information" is the content of the article: the news, knowledge, and information that people read newspapers for. There's also other information about the article itself that's important and helps the publisher: who wrote the article, where the summary of the article is, where the headline is, and similar data. And then there's "presentation": how the headline looks, the size of the print, where on the page the summary of the article appears, and other arrangements of the information. The *Journal* publishes the same information, more or less, in each edition, but the presentation of that information is different. The Web edition has hyperlinks, for example, and the mobile edition may leave out some information that's tagged as background.

To cut its expenses and get more flexibility, the *Journal* decided to disaggregate the task of writing the story from the task of presenting it in various formats. The writers would write the text just once and "tag" the data in the story so that computers could find it; computers would grab that story, look at the tags, and automatically reformat the story for the different editions.

To get this innovation to work, the *Journal* turned to a powerful revolutionary technology called eXtensible Markup Language (XML). XML provides a way to take text, such as the text of an news article, and put in "tags" that describe the text. XML also provides a way to define rules about how to manipulate the text based on the tags. With XML in place, the *Journal* process looks very different than before.

Reporters use an in-house, souped-up word processor to write a story. They highlight headlines, summary paragraphs, and similar information; behind the scenes, the computer captures the story in an XML document. When the story is finished, computers apply various sets of rules to the story to generate the different editions.

There are many benefits to this disaggregated system. Costs are much lower because it's not necessary to edit each edition. The *Journal*

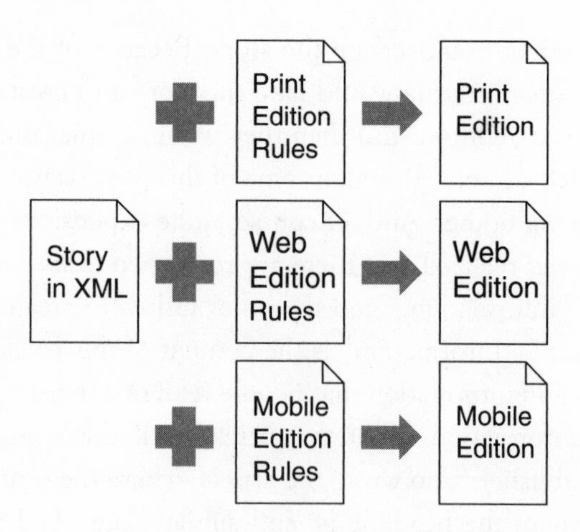

now has a flexibility it didn't have before—it can add a brand new way to publish its content, e.g., via instant messaging, by running the story through a new set of rules. The system also makes it easier to be creative. For example, to modify how the Web site looks, the *Journal* changes the rules—they don't have to go back and modify each individual web page.

The *Journal* is not alone; XML is an extremely popular technology, and it has revolutionized how computers handle documents. Computers use XML to send information to each other, store information for later use, or transform that information for people to see.

As we noted earlier in the book, there are a few indicators we can use to tell whether an innovation has the potential to sweep aside old ways of doing business. One indicator is whether there's a constant stream of follow-on innovations. XML certainly has follow-on innovations—XML is owned by the World Wide Web Consortium, and the World Wide Web Consortium follows the grand Internet tradition of continuous disaggregation. The result is a collection of related technologies that I'm grouping under the name XML. Another indication of the revolutionary potential is seeing how many disaggregations the innovation accomplishes—the more there are, the more power behind the avalanche. XML includes quite a few; here are some of the disaggregations included in XML's suite of technologies.

- *XML provides a standard way to tag data (information) in documents.* This is the basic disaggregation. In the old days, whenever someone came up with a new type of document, they also invented a new scheme to keep track of data in the document—and reinventing the wheel each time is very expensive. Now people just use XML instead.
- *XML tools let programmers read and write documents.* Because XML is standardized, all major programming languages have tools to read and write XML documents. To avoid the tedious chore of creating their own software, programmers incorporate these (disaggregated) tools into their own work.
- *XML provides tools to transform documents.* The rules to transform documents from one format to another (like the rules that the *Journal* uses to create different editions) can themselves be written in XML. That's another level of disaggregation that brings the usual benefits, including specialized tools devoted to document transformation.
- *XML disaggregates the job of validating documents.* XML provides a standard way to describe what a valid document looks like, and provides tools to check whether a document is valid.

Let me explain that last item and why it's very important. When a computer program reads or writes a document or sends it over a network, there's a chance that the document will accumulate some errors. Therefore, when a program reads a document, the program checks to make certain that the document is "valid"; for example, that no vital information is missing, and that all the data are in the right order. In the old days before XML, every time someone came up with a new type of document, they'd also have to come up with a program that could validate that new type of document—and writing those new programs was expensive and time-consuming.

XML provides a way to validate documents without custom programming. XML lets you write a "description" for each new type of document, and there are tools to check to see whether a document matches that description. When it comes time to read in a document, you take the document and the description and feed them both into your favorite XML validation tool. The job of checking a document against its description is disaggregated from the particular description.

XML is a quiet but very powerful revolutionary tool that has grown into a multibillion dollar industry of companies that provide XML tools and XML-based services. It's quiet because it's behind the scenes—XML hides in Web sites, inside databases, or in the documents computers send each other. Despite being out of the public eye, it's truly an avalanche: XML is sweeping away other methods of handling computer-driven documents.

The End of the Telephone Company: Internet Telephony

Why does your telephone ring when you're watching a football game? The telephone network is run by computers; logically, you ought to be able to program the computers not to bother you when the game is on. Wouldn't that be a great innovation, at least for football fanatics? I made up this hypothetical service, which I call the Home Team Football Network (HTFN), to illustrate how smart the telephone network could be. HTFN would know automatically when a game is on. If another HTFN fan calls during the game to discuss a play, the call will go through. Emergency calls always get through.

None of this is particularly complicated, but HTFN isn't available. Today's telephone companies don't seem to be interested in custom services like HTFN, and they have solid business and technical reasons to ignore HTFN. From a business point of view, not enough people would subscribe to HTFN to make it profitable for a company as large as a regional monopoly. From a technical point of view, telephone company computers aren't equipped to provide HTFN; the companies would have to invest money and effort. We won't see Home Team Football Network from today's telephone companies anytime soon.

If you like this hypothetical service or have ideas for one of your own, there is hope in a new telephone network that's beginning to replace today's telephone network. Instead of running over the old-style wires like the ones that today's phone companies use, this new network runs over the Internet. Better yet, this Internet-based network disaggregates today's telephone network—and disaggregation means universal benefits, such as creativity, with services like the Home Team Football Network not far behind.

This Internet-based service is usually called *Internet telephony,* or *IT* (this technology is also known as *Voice over IP* or *VoIP*).

Companies offer Internet telephony service today; small companies started the IT revolution, and then large companies jumped on the bandwagon. I happen to think some of these companies are going to lose their shirts, but first some details about Internet telephony.

Telephone calls traveling over what I call the "classic" telephone network require two basic functions from the telephone company. The first function is "routing": you dial a number, the telephone network figures out where to send your call and which wires to use, and the network sets up the call. The other basic function is "bandwidth," carrying your voice over the wires. For over a hundred years these two functions were bundled together, and you couldn't get one without the other. Whoever ran the wires to your home or put up the neighborhood's cellular phone tower "owned" your telephone call, soup to nuts—get one, get the other.

Internet telephony separates the two—IT disaggregates routing from bandwidth. Anyone, anywhere in the world who understands the technology can provide routing, and the bandwidth for the call comes from the same standard Internet connection that you use to download Web pages. In fact, that's a good way to think of it: Internet telephony service is like looking at Web pages. You can look at *anyone's* Web pages over your Internet connection, and IT lets *anyone* provide you with telephone service over your Internet connection.

How Web Pages Work	How Internet Telephony Works
Start: Run your Web browser program.	Start: Run your Internet telephony program.
Type in the address of a Web page; connect to the Web page "server" that has the Web page you want.	Enter the phone number you want; connect to the IT "server" which keeps track of other telephones.
The Web page "server" sends you a Web page. Data travels over your Internet connection.	The IT "server" connects your telephone to the other telephone. Data travels over your Internet connection.

Whether it's a Web page or your IT service, your Internet service provider carries the data back and forth. Your Internet service provider doesn't run all the Web page servers on the Internet—you can get Web pages from anywhere on the Internet. The same holds true for Internet telephony—you *should* be able to get IT service from anywhere on the Internet, unless your Internet service provider deliberately interferes.

The current business and technical model of telephone service is One Big Company, which doles out telephone numbers in your region, runs lines to your home, and handles your service; cellular companies also provide service and some competitive pressure, but no more than a handful of cellular companies exist in any one city. Governments tax telephone service like crazy, and regulatory agencies tell telephone companies how to run their service. The entire system assumes that telephone service is a scarce resource, costs lots of money to set up, requires careful control, and can actually be controlled by the government.

When making a telephone call becomes like downloading a Web page, all this will change. Anyone can set up an Internet telephony server, just like anyone can set up a Web server, so telephone service will no longer be a scarce resource. Telephone service will be difficult to tax because the IT service can come from anywhere in the world, and IT is all just Internet data anyway. Is the government going to examine each and every packet on the Internet to find the ones that contain conversations? The idea of a government body regulating IT telephone service is as laughable as a government body regulating Web pages—not that they won't try in the short term, of course. They might even succeed, to some extent.

The price of IT service should drop rather quickly toward zero. Right now your Internet service provider gives you an e-mail address and routes your e-mail as part of its standard package of Internet services; there's no particular reason why they should stop at that, and I fully expect them to give you an IT address and to route your IT calls, also as part of your Internet service. Why not? Internet telephony is another basic Internet service, just another kind of data packet traveling through the network. I also expect that Internet-based companies will find ways to give away IT service for free, the way Internet-based companies give e-mail service away for free.

Internet telephony is a real service; the technology is available at the time of this writing and is becoming more popular every day, and, all around the world, subscribers pay for IT service. How powerful is this IT innovation, which I claim is an avalanche? Recall that if an innovation falls into multiple categories, it's more likely to be a powerful avalanche; from the number of categories that IT disaggregates, it's clearly an innovation to be reckoned with. Here are IT's disaggregations, by category:

- *Authority.* The telephone companies lose their authority to set technical and business standards. Here's an example. Telephone numbers are a limited resource in the classic telephone network—that's one reason why the United States keeps adding area codes. Telephone numbers are doled out by the telephone companies for high fees. Internet telephony makes telephone numbers the same as e-mail addresses—unregulated, unlimited, imaginative, and free.
- *Ownership.* Phone companies limit "dial tone" services (e.g., call waiting) to those few services that they think will be profitable. Internet telephony lets anyone enter the market with whatever services they think will be profitable, or for that matter, whatever they think will be just plain fun.
- *Space/time.* Classical telephone companies impose geographical restrictions—if you live in Chicago, you must have a Chicago area code. With IT, there's no relationship between your location and your phone number—the Internet doesn't care where you live. My next-door neighbor used IT to install a second phone line, with a Jerusalem telephone number, although he lives here in Chicago. Now his family in Jerusalem can reach him for the price of a local call.
- *Conceptual.* The entire idea of a telephone company becomes obsolete. There's no need for a huge, bureaucratic, highly regulated company to provide telephone calls any more than there's a need for a huge, bureaucratic, highly regulated company to provide Web pages.

The potential benefits from these disaggregations are absolutely staggering. Here are a few of them:

- *Creativity.* Web pages continue to amaze me—there's always some new service or industry springing up online. Now imagine the same creativity applied to routing your phone calls!

- *Cost reduction.* One reason that even the classical telephone companies are moving toward IT is because the Internet uses bandwidth far more efficiently than the classical telephone network.
- *Synergy.* The IT servers that route the calls are part of the Internet, so the IT servers can enhance their routing decisions with all sorts of data from the Internet. If your desktop computer is on the Internet, and your daily calendar is on your desktop computer, then the IT service can look at your calendar and hold your calls when you're in a meeting. Expect a lot of other clever ways to incorporate Internet-based data into the routing of your calls, up to and including local weather conditions. ("Sorry, surf's up! Please leave a message at the tone.")
- *Competition.* My plain-vanilla office computer has an IT package loaded on it. I've set it up to route calls for a few hundred of my friends, not that I've gotten around to turning the service on just yet; I could route calls for several thousand people without any trouble. In other words, anyone who's interested and has Internet access can provide IT service.

Like other Internet protocols, the Internet telephony specification disaggregates each of its functions into even finer and finer pieces—another reason to expect great things from IT. As with any Internet protocol, people use the technology to create new services that aren't really related to the original purpose. I have a friend who uses IT to send music across the Internet to his girlfriend who lives in a different city—her loudspeakers are just another extension on his personal IT network.

I emphasize one crucial category of disaggregation:

- *Mechanics.* The classic telephone network requires telephones—special pieces of equipment that plug into the classical network. Internet telephony runs over the Internet, which means that IT telephones are just computer programs. You don't need a "telephone"—a special piece of hardware—to make a phone call.

Telephone calls without telephones! "Telephones" used for Internet telephony will be amazingly cheap because they're just software—they can be placed into any computer that's connected to the Internet, and, because they can be, they will be. Recalling the discussion we had a few chapters back about cameras, we note that once digital photography disaggregated cameras from film, cameras showed up in all sorts of unlikely places, such as in PDAs and cell phones. Now that IT disaggregates telephones from hardware, telephones will also start showing up in unlikely places—but they'll be more ubiquitous because they won't need any special hardware beyond a microphone and speaker, which are cheap already. (At the time of this writing, some handheld computers already incorporate IT phones and wireless Internet connections, a combination that spells trouble for the current cellular networks.)

The trick will be to figure out what to do with software-only phones, but the sky's the limit. Why shouldn't your refrigerator call you to let you know if food starts to spoil? Yes, that idea has been around for years, but with cheap wireless Internet connections in every home it's finally realistic. No sane person will run an Internet connection to their refrigerator—well, a few of *my* friends have, then again I'm not claiming they're particularly sane—but today no cables are needed to hook up to the network, just a wireless card that costs a few dollars. One day all high-end refrigerators will have wireless access; when they do, they'll call when there's a problem.

Given that IT is such a terrific revolutionary technology and is poised to trigger an avalanche of change, why am I so positive that some IT service providers are going to lose their shirts?

The problem is their business models. I don't think that a business model that's built on charging a monthly fee for IT service, or one that charges per call, is sustainable. Those fees are based on scarcity, and there is no scarcity—quite the opposite.

I don't dispute that there's an actual need to charge for IT service today. Most telephones out there are still classic telephones, and the classical telephone companies charge real money to route calls between the Internet and the classic network. If you want your Internet telephony phone to be accessible from the classic telephone network and vice versa, you have to rent a classic telephone network number. The classic telephone network will be around for a long time, and its politics of scarcity will let companies continue to charge for telephone service—at least for a while longer.

I also expect continued efforts by local governments to legislate taxes, regulations, and "public safety" rules for IT service—rules that could easily put today's local telephone monopolies securely in the driver's seat even as they too switch over to IT-based service. Expect to see open warfare at the government regulatory agencies as various telecommunications industry factions attempt to influence laws to their advantage.

Regardless of these challenges, IT is clearly sweeping old technology aside. All sides in the telecommunications battle are adopting IT—the current local monopolies, the large ISPs, and the upstarts who just operate IT servers. Internet telephony promises some terrific benefits beyond simple cost savings; with any luck, we'll see not just inexpensive telephone service but entirely new definitions of what telephone service means.

The Amazing Exploding Computer: Jini and Bluetooth, or Something Like Them

Major manufacturers are at war to own my address book. Here's a list of the combatants.

- Cell phone companies want my address book in their cell phones.
- Manufacturers of PDAs want my address book in their devices but are willing to let me see it on their companion desktop applications.
- My desktop computer's e-mail program stores not only e-mail addresses, but also will let me store complete contact information.
- My laptop computer's e-mail program does the same thing as the desktop computer's program, but the two programs are incompatible and don't share their address book information.
- My local telephone company offers me a speed-dial list—and for a monthly fee they'll store my phone list on their network.
- My long-distance service provider offers a speed-dial list I can use when I call their toll-free number to make calls.
- My office telephone has a speed-dial list.
- My home telephones have speed-dial lists, too, but every room in the house has a different list.
- Business-networking Web sites let you add contact information.
- Web-based address books also let you add contact information.
- The "cell phone-in-a-car" concierge services also want my contact information.

And so on—I don't think I've actually covered them all; I didn't include "contact relationship management" programs on my desktop, or project management software. All of these programs want control over my address book.

What *I* want is fairly straightforward: I want my addresses and telephone numbers with me when I need them. I want them to be stored safely, I want to be able to choose how I access these names and numbers, I want to make certain no one else gets their hands on them without my permission, and I don't want to have to reenter all the information every time I buy a new PDA or cell phone, or switch telephone service. I want my address book available on whatever phone I'm using, no matter if I'm calling from my office, home, cell phone, car, or desktop computer, or for that matter even from a hotel phone. And I especially don't want to have to synchronize each and every change to my address book by reentering it by hand on a half-dozen different gizmos—for goodness sake, that's what computers were invented for!

Is there a way for me to get what I want? Can I have just one list that works everywhere? I think there is a way, and, to demonstrate, I'll discuss an innovation that I call the *exploding desktop computer*. The exploding desktop computer is based on two technologies, Bluetooth and Jini—or something like them.

For a while, Bluetooth technology was a solution in search of a problem. Bluetooth provides very short-range wireless networking, a great idea, but Bluetooth works at relatively low speeds and never did manage to replace cables as originally intended, e.g., to connect disk drives to computers. Bluetooth got a second chance when it emerged as a decent way to make low-speed connections. My cell phone is equipped with Bluetooth, and instead of using a headset with wires I can use a wireless Bluetooth headset. I also have Bluetooth on my laptop computer, and my cell phone connects to the laptop using Bluetooth—I use the laptop's full-sized keyboard and screen to edit the phone's address book. Bluetooth, or rather something like it that works at higher speeds and has better security, constitutes one member of a pair of revolutionary technologies.

Jini is the other member of the pair. Jini is a computer protocol that lets a device (a printer, a disk drive, a monitor) advertise to other devices what it's good for: "I'm a printer, I can print in color." "I'm a keyboard, English-language." Jini lets you hook the devices to each other, automatically, to form a working collection of computer parts.

Put Jini and Bluetooth together and I predict that they, or something like them, will disaggregate the personal computer. That's because current personal computers and electronic gear—despite the fact that we've grown used to them and we can get along using them—don't really do what we want them to do. When I'm on the road, I have a laptop computer, a PDA, and a cell phone. But I don't really *want* any of these things. What I really want is at least two different sizes of display screens (pocket size and full size), access to all my files, a few decent ways to enter data, and Internet/telephone/fax connectivity. Bluetooth and Jini, or something like them, can give me what I really want if these two technologies disaggregate the personal computer, the PDA, and the cell phone down to their constituent parts.

There are actually two paths for this revolution. On one path, we take the desktop computer, the PDA, and the cell phone and blow them all to pieces, hence the idea of the exploding desktop computer. The keyboard, CPU, disk drive, mouse, and display all become separate parts. Instead of a PDA, I carry a small display screen that also accepts pen or spoken input. My cell phone morphs into a headset with microphone, and, if there's a keypad—if I don't just use speech recognition to dial—the keypad is probably on the screen of PDA. All my files are in the same place—on the disaggregated disk drive—and how I access the files depends on what I'm doing. If I'm in my hotel room, I can unpack my bags and get out my big display, full-sized keyboard, and mouse. Bluetooth wires them together, Jini creates a working collection, and I start typing away on what looks like a traditional computer. If I'm in a taxi, at the airport, or at a meeting, I can use the pocket-sized screen and pen to get at my files, which are on the disk drive in my attaché case. Through the magic of Jini and Bluetooth, at any given time my computer lets me use the most convenient input and output devices for the situation. I go from an ad-hoc collection of electronic gear to a smoothly integrated machine.

There's another possible path for this revolution. Don't disaggregate the hardware; disaggregate the tasks the computer does, and then build a separate device for each task—something I'll call a "task device." A good example of a task device is one for the lowly address book—the list of names, telephone numbers, and e-mail addresses that I carry everywhere I go. I've already discussed the battle of the address book and how various industries want to treat my address book information.

The Bluetooth/Jini revolution can solve this by creating a tiny black box, let's say something that fits on my key chain, that contains my address book—a task device to store and manage my address book information. Bluetooth will make it visible to any authorized computer, and Jini will advertise what the task device can do. This task device disaggregates the "address book" function from a wide range of electronic gear and brings it into a common database. Any smart phone and any smart computer program will use the address book task device to get my names and telephone numbers and make it available—to me only.

The same idea can be extended to other tasks. What about my date book? Document storage? Word processing? Playing music? Web browsing? In each case, the same information and the same task are done by many different machines. I would like a task device to disaggregate my music collection and make it accessible from any location—on dedicated players, on cell phones, on my home stereo, or in my car.

The "exploding desktop computer" is already available to some extent. Companies now offer "network storage" for the home market, as they have for years in the business market, that is, disk drives that reside on the computer network and provide storage to all computers on your network. Network printers offer print services to any computer on your network—this innovation disaggregates the printer from any one computer. Recently I've heard about networked monitors—a monitor that can be used by any computer on the network.

As for task devices, I see indications that task devices may one day come in some form or another. From time to time, the industry takes a few small steps in the right direction. Entries from an address book can be put into a disaggregated "vCard" format that's understood by many different software programs; that's how I get address information out of my PDA and into my cell phone. Some companies specialize in software to synchronize the address book of a PDA with address information on desktop computers. These are steps toward an address book task device. The address book task device would be quite convenient, but will it ever be built? Only time will tell.

Chapter Thirteen

Getting Started, Finishing Touches

How to Start

If you have a problem in hand and want a solution, here's how to get started. Even if you don't have a burning problem to solve, I recommend that you pick one anyway and apply disaggregation to solve it by following the steps in this chapter. (I'm a big believer in practical work to make certain that I understand what I've studied.) In the rest of this section, I'll describe the steps to take to turn a problem into a solution.

Ask the Right Question

Chapter 4 presents the basic framework of how to use disaggregation to solve problems. We'll start with the steps outlined in Chapter

4 (devise, interface, accept, and evaluate) and add material from other chapters as we go along.

The first step is to see whether you find a solution to your problem immediately. A plain, uncomplicated solution may just leap out at you, especially after practice with the material in this book. If nothing is immediately obvious, here are three methods for generating an innovation:

- Restate the problem in terms of disaggregation or the benefits of disaggregation. Chapter 3 introduces the benefits of disaggregation, and the case studies in Part II discuss the benefits of particular innovations in great detail.
- Think about how the problem relates to the five categories of disaggregation. Chapter 2 explains the categories.
- Brainstorm by taking each and every piece of the existing infrastructure apart—the "smash and grab" method.

Once you've formulated an innovation that solves your problem, record your ideas and continue to the next step.

Assess and Extend the Proposal

In Chapter 2, we saw how the most powerful avalanches start: when the innovation disaggregates in multiple categories, when it has many disaggregations even in a single category, or when the innovation fulfills a basic human desire in a particularly effective way. Can you improve your innovation to make it—potentially at least—more powerful?

Start by assessing your innovation. It's a disaggregation—what categories does it disaggregate? What basic human desires does it fulfill? Go through each category—they're all listed in the worksheet shown in Table 13.1—and write down what, if anything, your innovation does in that category. Sometimes an innovation will fall completely into one single category; sometimes it will hit a few categories. Don't forget that an innovation can have several entries in a single category, as we saw with the automobile in Chapter 6.

Next, go through each of the benefits in the worksheet and decide whether your innovation will deliver that benefit. Again, don't be surprised to see multiple entries under a single benefit or no entries at all under a different one.

Table 13.1 Innovation Worksheet

Innovation Description: ______________

How It Works: ______________________

CATEGORIES:

Authority	
Ownership	
Mechanics	
Space/Time	
Concepts	

BENEFITS:

Creativity	
Cost Reduction	
Competition	
Simplicity	
Specialization	
Synergy	

Once you've filled out the worksheet (Table 13.1), it's time to think about ways to fill in any remaining blank spaces. Are there ways to extend your innovation?

- If it does not disaggregate in one of the categories, is there a way to reformulate the innovation to include that category?
- Is some benefit missing? If so, can you provide that benefit if you modify your innovation?
- Can the innovation extend into other, related areas? This isn't filling in the blanks as much as it is extending the scope—the more disaggregations, the merrier (and the more powerful).

Although you don't want to bite off more than you can chew, aim as high as you can—see how far your innovation can stretch. You can't go wrong if you're as ambitious as possible and then cut back your proposal to something manageable. That's because revolutions tend to generate new revolutions, as we see throughout this book; it pays to push your idea to its limits to help you understand the big picture and where your innovation may lead someday. You need a clear idea of what you can accomplish in practice, but a roadmap of future developments is also useful.

Create Interfaces

Disaggregation takes things apart—things that work and provide useful functions. The pieces that you create through disaggregation need interfaces to work together afterward; otherwise the useful function you started with may be lost.

Depending on the innovation, the interfaces can be public or private. Chapter 4 discusses public and private interfaces and when it's appropriate to use one or the other.

Generate Acceptance

Unless you're a one-person business and no one else will ever see your innovation, you must market your innovation. Your target audience consists of members of your own company; your business partners; your industry peers; and, finally, your customers, who ultimately decide whether an innovation thrives. If your industry peers can't figure out how to cooperate, everyone will suffer; Chapter 4 provides examples of disastrous industry turf wars.

Standards are often a good way to gain trust. Check the criteria in Chapter 4 to the determine the role of standards for your innovation. Here are some possibilities:

- Standards might make your innovation more acceptable to your company, your business partners, industry peers, and customers.
- You may find that your innovation would be most acceptable if it became an official standard.
- You may find that standards are irrelevant to your innovation, especially if your interfaces are completely private.

As we saw in Chapter 8, without standards much of modern civilization just wouldn't function—in some industries you simply must standardize your interfaces if you want anyone to take your innovation seriously.

Consider how companies who don't accept your innovation will react. Companies may simply reject your initiative or they may actively oppose it. Chapter 10 includes counterrevolutionary tactics that, thankfully, aren't very common. Chapter 11 discusses the role of government; if your innovation disturbs established companies with comfortable political connections, your battle may spill over into legislative chambers and courtrooms.

Execute

Although I've recorded all these steps—generating the innovation, creating interfaces, selling your innovation—one after the other, in practice they proceed in parallel. As you generate your innovation, you'll want to look forward to the interfaces you'll need; as you discuss the innovation with potential partners, you'll find yourself modifying your original proposal. When you mix these steps together, it's not a symptom of confusion—instead, taking the steps in parallel lets you apply feedback to improve your innovation.

If you're just doing a pen-and-paper exercise, at this point you've gone about as far as you can go. The next step is to actually build the innovation. Building the innovation usually happens somewhat in parallel with the previous steps.

Evaluate

Once your innovation is real and your product is rolling out the door, certainly it's time to celebrate and enjoy your victory. The next step, as outlined in Chapter 4, is to roll up your sleeves and evaluate what happened. Did you see the benefits you projected? What unexpected benefits did you see? What did your competitors do?

Most of all, what new innovations can you incorporate into your next round of products? You've triggered an avalanche, and now's the time to enjoy the ride.

Finishing Touches: The Thoughts That Got Away

A great deal of material never made it into the final version of this book. The principles in this book apply to charitable organizations—

innovation is just as crucial in volunteer work as it is anywhere else—but since the principles are the same, they aren't discussed separately. A friend who works in medicine explained how disaggregation helped advance basic areas of medical research; that was wandering too far afield. In Western civilizations, the power of the monarchy disaggregated into separate judicial, legislative, and executive branches of government, with church outside the government entirely. That idea and other broad philosophical questions about society provide interesting topics to explore and understand in terms of disaggregation; again, they're off the main topics of business and technology.

Every day my newspaper brings me at least one new thought—another innovation to think about. Innovations that are likely to trigger a sweeping avalanche are somewhat rare, but innovations based on revolutionary technology pop up all the time—innovations that I instantly recognize as benefits of disaggregation.

To keep track of new innovations, to present some of the information that didn't make it into the book, and to foster new ideas, I've decided to conduct a continuing conversation, and this is your invitation to join in. This book's Web site is at http://www.PebbleAndAvalanche.com, and there you'll find news, articles, a mailing list, and further information. You will also find a larger version of the worksheet I included in this chapter.

True to the principles of this book, I've disaggregated the function of managing the Web site—that's my job—from the function of writing the material that appears there. You can contribute material to the site if you like. Visitors will be able expand, expound, and explain the issues. I look forward to seeing you there.

Endnotes

Chapter 2. Starting Revolutions: What to Take Apart

11 **Patient satisfaction improves when patients exercise autonomy in their health care decisions:** M. Gattellari, P. N. Butow, and M. H. Tattersall, "Sharing decisions in cancer care," *Social Science & Medicine,* 52(12):1865–1878. Note that the effect of ethics committees is not measured in these studies, just the correlation between patient satisfaction and patient autonomy.

14 **Henry Ford and his team were among the first people to use electric motors to power factory tools:** Peter J. Ling, *America and the Automobile: Technology, Reform, and Social Change,* Manchester, U.K.: Manchester University Press, 1990, p. 141. For a review of mechanical transmissions in factories, see Austin Weber, "Line Shafts and Belts," Assembly Magazine, October 1, 2003, http://www.assemblymag.com/CDA/ArticleInformation/features/BNP__Features__Item/0,6493,110490,00.html. Contrary to popular belief, Ford did not invent the production line; Ford said he got the idea from the meatpacking industry, which disassembled animal carcasses by moving them along lines in the 1800s. Ford was the first to use moving conveyers in the automotive industry. John B. Rae, *The*

American Automobile Industry, Boston: Twayne Publishers, 1984, pp. 35–39. See also Robert Paul Thomas, *An Analysis of the Pattern of Growth of the Automobile Industry 1895-1929,* New York: Arno Press, 1977, p. 136.

15 **Money is just an arbitrary token that people use to keep track of the value of other things:** This description applies regardless of whether U.S. currency is a "credit money" or a "fiat money." Ludwig von Mises, *Human Action: A Treatise on Economics,* New Haven, CT: Yale University Press, 1963, pp. 428–429.

Chapter 4. Four Stages to Revolution: Devise, Interface, Accept, Evaluate

29 **I wonder whether it's time to reconfigure the Web site's computer system to handle text and graphics separately:** A typical solution to that particular problem.

34 **eBay disaggregated its internal computer system:** eBay hasn't revealed full details of their architecture to the general public, and the text is based on reading between the lines of publicly available information. The actual sequence of events and the details of the architecture will of course be more complex. As they rolled out their application programming interface in 2000, eBay introduced the "disaggregation" (in their words) of their internal architecture to increase reliability and offer new services. Stan Gibson, "eBay: Sold on Grid," *Oracle,* August, 2004, http://www.findarticles.com/p/articles/mi_zdora/is_200408/ai_n7184451. For security and ease of development, eBay disaggregates access to their servers even for their internal developers. Sanjay M. Krishnamurthy, "The eBay Global Platform and Oracle 10g JDBC: Scaling for Future Growth, 2004, http://download-east.oracle.com/oowsf2004/1235_wp.pdf.

35 **Recently some 40% of eBay's listings were generated through these software packages:** Will Iverson, "Web Services in Action: Integrating with the eBay® Marketplace," June 2004, O'Reilly Media, white paper, p. 7.

36 **A crucial part of eBay's strategy was to use standards in its public interfaces:** Iverson, p. 8. See also Pat Martin, "WebSphere Beats the Competition," *WebSphere Journal,* http://websphere.sys-con.com/read/43063.htm.

Chapter 5. From Horses and Buggies to Jet Planes: The Revolution in Manufacturing

44 **Before Whitney, each gun, like all manufactured goods of that era, was a unique product; craftsmen painstakingly created each gun individually:**

Constance M. Green, *Eli Whitney and the Birth of American Technology* Boston: Little, Brown and Company, 1956, pp. 103–104. Whitney did not claim to have invented the system of interchangeable parts. For a dissenting view on whether Whitney should be credited, see David A. Hounshell, *From the American System to Mass Production 1800-1932: The Development of Manufacturing in the United States,* Baltimore: The Johns Hopkins University Press, 1984, pp. 30–31. This controversy is over whether Whitney should be credited with the discovery, not the facts of how well Whitney's system works.

45 **Whitney is the inventor of the modern tool-and-die industry:** Green, p. 107, notes the pioneering ideas of Simeon North. Whitney's work and Whitney's success certainly captured the imagination of manufacturers around the world. The French did introduce some interchangeable parts in 1765, but they used craft methods. Hounshell, pp. 25–28.

45 **Four thousand clocks was an absolutely staggering number, more than a traditional clockmaker would make in a lifetime, much less three years!:** Donald R. Hoke, *Ingenious Yankees: The Rise of the American System of Manufactures in the Private Sector,* New York: Columbia University Press, 1990, p. 53.

46 **The manufacturers invented clever and inexpensive ways to measure their products to make certain they met the standards:** Hoke, pp. 74, 88.

47 **When Terry started his clock business clocks cost about $40 retail—there were no clock wholesalers because the concept did not exist:** Hoke, pp. 56–59.

48 **but in a curiously collegial way:** Hoke, pp. 62–63.

48 **Records survive:** Hoke, p. 64.

Chapter 6. The Automobile Takes On the Railroads

52 **The question of which European country had the right to build a railway in an Eastern or African colony nearly started a war or two:** In particular, I am thinking of the heightened tensions in the Belgian Congo between Belgium and Germany over rights-of-way in the late 1800s and early 1900s.

54 **if the railroad is run by the government, the decision requires high-level political and economic maneuvering because of the high costs:** In the early days of railroads, some communities built railroad branches to connect their towns to the national network.

56 **smugly ignored the advice of the remaining Southern Pacific personnel:** "The surviving S[outhern] P[acific] types tried to explain what happened the last time that particular cut was tried, but no one was listening." Guy Span, "The Great Union Pacific Railroad Service Meltdown: Could It Happen Again?", San Francisco BayCrossings.com, http://www.baycrossings.com/Archives/2004/05_June/the_great_ union_pacific_railroad_meltdown.htm.

Chapter 7. The Internet's Permanent Revolution

62 **ordinary computer underneath someone's office desk:** The root servers are far more closely guarded today.

62 **There might be up to ten computer networks in each country to carry data:** The United States obtained special permission to have a two-digit country code, which allowed up to 100 networks. According to author Janet Abbate (*Inventing the Internet,* Cambridge, MA: The MIT Press, 1999, p 166), the United States could have up to 200 networks, but it's not clear how she arrived at that figure. The ITU currently lists 70 public data networks for the United States. International Telecommunication Union / Telecommunication Standardization Bureau of ITU, List of Data Network Identification Codes (DNIC) (According to ITU-T Recommendation X.121) (Position on 15 August 2004 [10/00]), Annex to ITU Operational Bulletin No. 818.

64 **A committee of the ITU eventually produced the X.25 specification for data networks:** For an excellent review of X.25, see Abbate, *Inventing the Internet,* pp. 152–167. Abbate also reviews SNA and OSI, other serious alternatives to the Internet.

66 **Despite all this, the switches continued to work and calls went through:** Eventually the loss of electrical power and rising water forced the equipment out of service, and the damage did require massive rebuilding, which disrupted service all throughout Manhattan and indeed the world. But the equipment itself was impervious to shocks that would have taken lesser equipment completely out of service. Thomas D. O'Rourke, Arthur J. Lembo, and Linda K. Nozick, "Lessons Learned from the World Trade Center Disaster About Critical Utility Systems," in *Beyond September 11th: An Account of Post-Disaster Research*, ISBN 1-877943-16-9, published by the National Hazards Center of University of Colorado at Boulder, p. 277, http://www.colorado.edu/hazards/sp/sp39/sept11book_ch10_orourke.pdf. Personnel at the facility deliberately turned off the equipment. Because they had to stand in water to reach the equipment, they used wooden poles to avoid electrical shocks. Glenn Bischoff, "Recovery," Telephony Online, Nov 19, 2001, http://telephonyonline.com/mag/telecom_recovery_2. The switches still functioned; see interview with Verizon's co-chairman, "Communication in a Time of Tragedy," *Ability Magazine,* http://ability-magazine.com/nichols_verizon.html.

66 **And another telephone switch *in the basement* of one of the collapsed World Trade Center towers was still working the next day!:** A Reuters wire service report quotes AT&T spokesman Dave Johnson. "AT&T Equipment Survived Trade Center Collapse," Sept 12, 2001. See, for example, http://www.bellsystemmemorial. com/ miscellaneous.html#AT&T%20equipment%20survived%20trade%20center%20collapse. I was skeptical of this story because I believed that everything in basements would have been crushed flat by the collapse of the towers, but some sections of the

multiple basements under the World Trade Center did survive. George J. Tamaro, "World Trade Center 'Bathtub': From Genesis to Armageddon," *The Bridge,* 32(1), http://www.nae.edu/nae/bridgecom.nsf/weblinks/CGOZ-58NLJ9?OpenDocument.

69 **the flexibility of Internet specifications:** New Internet technologies, such as file sharing programs (e.g., BitTorrent) and broadcast television over the Internet, rely on new data formats and clever use of the envelopes—and *the ability to experiment with the network itself,* which is implicit in how the Internet operates.

70 **192.0.34.166:** Actually, that's the address of www.example.com. I'm not crazy enough to publish the IP address of my office network.

71 **Routers for small local networks can be inexpensive and "dumb" because they only confront a limited set of problems:** At the national level, routers are amazingly brilliant.

74 **Internet-based e-mail started out as a "hack" put together out of existing systems:** Katie Hafner and Matthew Lyon, Talking Headers, (This excerpt of Wizards, a history of e-mail titled "Talking Headers," appeared in the *Washington Post Magazine* on August 4, 1996. It was edited by Bob Thompson and John Cotter.) http://www.olografix.org/gubi/estate/libri/wizards/email.html.

Chapter 8. Interfaces and Standards: The Nuts and Bolts of Modern Civilization

83 **Home exteriors can be repaired because masonry bricks and cinder blocks come in the same sizes they always have:** Lumber changed sizes, however, especially in the 1970s. A 2 by 4 was two actually inches by four inches years ago, and is now $1\frac{1}{2}$ inches by $3\frac{1}{2}$ inches. This causes problems for people who repair older homes.

83 **The people who introduced the idea of standardization for all the "nuts and bolts" of modern civilization were, ironically, the manufacturers of nuts and bolts:** An excellent article on this subject, which first brought this example of disaggregation to my attention: James Surowiecki, "Turn of the Century," *Wired,* January 2002, p 85.

84 **fix a bicycle:** In an interesting counterexample to the adoption of standard threads, Raleigh bicycles used proprietary threads until the 1990s. Sheldon "Nottingham" Brown, "Threading/interchangeability Issues for Older Raleigh Bicycles," 2002, http://www.sheldonbrown.com/raleigh26.html.

85 **the local blacksmith could make something that fit:** Tom Kelleher, "Nuts and bolts," September 2002, *The Chronicle of the Early American Industries Association, Inc.,* http://www.findarticles.com/p/articles/mi_qa3983/is_200209/ ai_n9102183/ pg_2.

85 **In today's money, that's about $600!:** A decent comparison of prices is very difficult when dealing with two very different eras. The $600 figure comes from a comparison of consumer price indexes. If I were to use instead the number of

hours of unskilled labor needed to purchase a clock, the price is closer to $10,000. See the calculator and explanations at Economic History Association, "What Is Its Relative Value in US Dollars?," http://eh.net/hmit/compare.

86 **a long debate about the merits of standards:** Bruce Sinclair, "At the Turn of a Screw: William Sellers, the Franklin Institute, and a Standard American Thread," *Technology and Culture,* 10(1), p. 27. As Sinclair puts it, the invention is only half the battle.

87 **any move to eliminate differences also eliminates a competitive advantage:** Sinclair, p. 20, notes that screw manufacturers deliberately made nonstandard threads to lock in their customers.

88 **that famous engineering malady known as NIH:** Sinclair, p. 31, shows that these squabbles reached international status as the United States, England, and Europe failed to converge on a single standard.

91 **hundreds of thousands of standards:** The "NSSN" project is a joint effort by the American National Standards Institute and other agencies. The site provides information on over 250,000 standards. http://www.nssn.org/index.html.

91 **there was a battle in progress on the shelves of your local computer store:** For a current summary of technology differences and a succinct history of the different formats, see Jim Taylor, "DVD Frequently Asked Questions," DVD Demystified, http://www.dvddemystified.com/dvdfaq.html.

92 **but *that* standard is in dispute as well:** David Carnoy, "HD-DVD vs. Blu-ray: Who Cares?", CNET.com, December 14, 2004, http://www.cnet.com.au/desktops/storage/0,39029473,40003001,00.htm.

92 **They're fighting over something near and dear to their hearts: money:** A paraphrase of a statement made, in an entirely different context, by satirist Tom Lehrer.

Chapter 9. Coping with Surprises

101 **Kodak did have some interesting things going for it when the revolution hit:** Robert M. Grant, *Cases to Accompany Contemporary Strategy Analysis,* 2005, Oxford, U.K.: Blackwell Publishing, Chapter 6.

102 **year-by-year list of corporate milestones:** "History of Kodak: Corporate Milestones 1990–1999," http://www.kodak.com/US/en/corp/kodakHistory/1990_1999.shtml.

103 **"take the free [available] cash and set it on fire":** Comment by "buffetstudent," "What Buffett Might Say about EK," Motley Fool, September 26, 2003, http://netscape.fool.com/community/pod/2003/030926.htm. A cursory search of news items of the time will reveal the comments of many other disenchanted shareholders.

Chapter 10. Marx, Lenin, and Gates: Failed Counterrevolutions

108 **tens of millions of its own citizens:** Estimate vary widely for the number of people killed during the Stalin era by starvation, in purges, and in labor camps. Numbers range from a "low" of twenty million to as high as sixty million or more. These estimates include well over ten million people that starved to death during the 1920s and 1930s as a result of ineptitude and/or deliberate policy.

108 **costs several billion dollars:** For example, "Powerchip to build two 12-inch memory fabs," *The Inquirer,* November 16, 2004, http://www.theinquirer.net/default.aspx?article=19699. For a projection of rising costs, see "Can the Semiconductor Industry Afford the Cost of New Fabs?", IC Knowledge LLC, http://www.icknowledge.com/economics/fab_costs.html.

109 **about 10% of the U.S. money supply consists of worthless pieces of scrap paper:** Rough estimate based on 2004 U.S. money supply figures as published by the Federal Reserve Board. The United States has relatively high amounts of paper money per capita, probably owing to the popularity of U.S. currency in the global underground economy.

111 **With price's information gone, the interface to allocate goods and services was gone:** The "calculation problem." Ludwig von Mises, *Human Action: A Treatise on Human Economics,* New Haven, CT: Yale University Press, 1963, p. 698ff.

111 **another important job goal of managers was to avoid being shot:** Paul R. Gregory, *The Political Economy of Stalinism: Evidence from the Soviet Secret Archives,* Cambridge U.K., Cambridge University Press, 2004, pp. 121, 170. Books on the economics of communist countries seem to always include the key words "executions" and "purges" in their indexes.

111 **plenty of shoes each year, three pairs for every person:** Scott Shane, *Dismantling Utopia: How Information Ended the Soviet Union,* Chicago: Ivan R. Dee, 1994, pp. 77–78.

112 **no more than half of the nuts and bolts were made in the official factories:** The "autarchy" problem. David Granick, *The Red Executive: A Study of the Organization Man in Russian Industry,* London: Macmillan, 1960, pp. 160–161.

112 **No disaggregation means no benefits of disaggregation:** Granick, Red Executive, pp. 247–249. Granick ascribes quality control problems to lack of expensive quality-control equipment in the duplicate facilities.

113 **why does Microsoft act like a bully:** See, for example, Microsoft's dealings with both Go and Vobis. Wendy Goldman Rohm, *The Microsoft File: The Secret Case Against Bill Gates,* New York: Times Business, 1998, pp. 94–96.

113 **Microsoft has an aggressive internal culture:** See, for example, "What Penalties for Microsoft," *BusinessWeek Online,* February 22, 1999, http://www.businessweek.com/1999/99_08/b3617027.htm

114 **change something in the operating system to make applications from other companies fail:** The most famous case of this was not in the fundamental operating system, but the Windows "presentation manager," which ran on top of the DOS operating system. (At that time, DOS and Windows were separate; MS has since reaggregated the operating system and presentation manager.) Copies of Windows that users ran on top of a competing operating system, DR-DOS, would issue dire warnings. See *The Microsoft File,* pp. 116–117, 236. Another notorious case is the purported sabotage of Lotus 1-2-3, the rival to Microsoft's Multiplan, with the slogan "DOS isn't done until Lotus won't run." James Wallace and Jim Erickson, *Hard Drive: Bill Gates and the Making of the Microsoft Empire,* New York: John Wiley & Sons, Inc., 1992, p. 233. These accusations have not been repeated in recent years.

114 **Microsoft was accused of this for years:** Mike Ricciuti, "Platform Ploys in the Public Eye," December 2, 1998, CNET News.com, http://news.com.com/Platform+ploys+in+the+public+eye+page+2/2009-1001_3-218437-2.html?tag=st.next. I have not heard these accusations in recent years, possibly because of the impact of antitrust scrutiny on Microsoft's behavior.

114 **while other software vendors scramble to fix their software:** "Microsoft's interactions with Netscape, IBM, Intel, Apple, and RealNetworks all reveal Microsoft's business strategy of directing its monopoly power toward inducing other companies to abandon projects that threaten Microsoft and toward punishing those companies that resist." Judge Jackson. U.S. District Court Judge Thomas Penfield Jackson, U.S. v. Microsoft: Court's Finding of Fact, November 5, 1999, http://www.usdoj.gov/atr/cases/f3800/msjudgex.htm, paragraph 132. For a specific example of delaying a competitor's product release, see paragraphs 90–92.

114 **the "hidden API" tactic, where Microsoft keeps some of the API secret:** Some of these hidden APIs have been reverse-engineered and published in books. For an article, see Bruce Ediger, "Windows NT, Secret APIs, and the Consequences," http://www.users.qwest.net/~eballen1/nt.sekrits.html. The recent EU antitrust decision requires Microsoft to make its APIs more publicly available. David Worthington, BetaNews, "Microsoft Accepts Most EU Demands," April 4, 2005, http://www.betanews.com/article/Microsoft_Accepts_Most_EU_Demands/1112657252.

116 **Internet Explorer would start to use proprietary technology to interact with Web servers:** Microsoft writes, "By extending these protocols and developing new protocols, we can deny OSS [Open Source Software] projects entry into the market." Eric S. Raymond (editor), "Halloween Documents I," version 1.14, http://www.opensource.org/halloween/halloween1.php.

116 **Microsoft tried this against the Java programming language, which they quite correctly viewed as a threat:** Judge Jackson, Court Findings, Section IV B. For viewing as a threat, see "Halloween Documents II," http://www.opensource .org/halloween/halloween2.php.

117 **Microsoft paid Sun about $2 billion dollars to settle various antitrust and patent claims:** Paul R. LaMonica, CNN Money, "Bill and Scott: Let's Be Friends," CNN/Money, April 2, 2004, http://money.cnn.com/2004/04/02/technology/microsoftsun/?cnn=yes.

117 **Linux dominates the Internet infrastructure:** In Web servers, for example, see "April 2005 Web Server Survey," Netcraft, http://news.netcraft.com/archives/2005/04/01/april_2005_web_server_survey.html.

117 **Linux continues to make inroads into the back-office operations of businesses:** "Unix servers up 2.7%, Linux servers up 35.6%," *IT Fact,* February 28, 2005, http://www.itfacts.biz/index.php?id=P2712.

118 **With thousands of "eyeballs" looking for errors:** Also expressed as, "Given enough eyeballs, all bugs are shallow." Eric S. Raymond, *The Cathedral and the Bazaar,* Version 3.0, http://www.catb.org/~esr/writings/cathedral-bazaar/cathedral-bazaar/ar01s04.html. *The Cathedral and the Bazaar* is also available as a book from O'Reilly.

119 **critics and governmental authorities denounced Microsoft advertisements as misleading:** Government authorities: "Microsoft slammed over misleading Windows Linux claims," *The Inquirer,* August 25, 2004, http://www.theinquirer.net/?article=18067. Other criticism: Eric S. Raymond, "Get the FUD," June 22, 2004, http://www.opensource.org/halloween/halloween11.html.

Chapter 11. The Role of Government

121 **Old Peculier beer:** Not a typographical error; *peculier* as in "special." An ale, brewed since the 1800s in Masham, England.

122 **beer sales in Pennsylvania:** Joe Sixpack, "Finally, Sunday sales starting at State Stores," Philly.com, February 7, 2003, http://www.ledger-enquirer.com/mld/philly/entertainment/columnists/joe_sixpack/5127098.htm. My thanks to Joe Sixpack for a private communication on the topic.

123 **rather thin excuse:** Robert D. Atkinson, "Revenge of the Disintermediated: How the Middleman is Fighting E-Commerce and Hurting Consumers," policy report, Progressive Policy Institute, January 2001, http://www.ppionline.org/documents/disintermediated.pdf, p 9. Atkinson notes that a requirement that deliveries be made only to people over 21 would eliminate this objection, but lawmakers ignore that solution.

123 **According to the U.S. Federal Trade Commission, direct automobile sales would cut about 10% off the price of a new car:** "Prepared Statement of the Federal Trade Commission Before the Subcommittee on Commerce, Trade, and Consumer Protection Committee on Energy and Commerce, United States House of Representatives, Washington, D.C., September 26, 2002," http://www.ftc.gov/os/ 2002/09/020926testimony.htm. The FTC quotes, with

apparent approval, information provided by the Consumer Federation of America. Another cited study shows that women and minorities, who usually pay more for sales conducted in person, pay less when they purchase cars over the Internet.

123 **In response to pressure from the automobile dealers, just about every state in the United States amended their franchise laws to restrict automobile sales over the Internet:** Atkinson, "Revenge," p. 7. By 2005, just about every state restricted new car sales over the Internet.

124 **Apparently the industry couldn't muster an argument they could repeat with a straight face in court:** One court noted that a coffin is "nothing more than a glorified box." Quoted in John R. Wilke, "Funeral Industry is Hit with a Casket-Pricing Suit," *The Wall Street Journal,* May 4, 2005, p. D1.

126 **in 2004 spent an estimated $18 million:** Rick Anderson, "How to Excel in D.C.: Spread Lots of Money Around," *Seattle Weekly,* September 22–28, 2004, http://www.seattleweekly.com/features/0438/040922_news_microsoft.php.

126 **a coppersmith in Nuremberg invented an improved metal lathe:** L. Sprague de Camp, *The Ancient Engineers,* New York: Ballantine Books, 1974, p. 367.

Chapter 12. Predictions: Three Revolutions in Progress

128 **an RSS news feed:** RSS stands for "Really Simple Syndication," a method to send brief announcements of news and information over the Internet. The meaning of the acronym has changed over the years.

129 **the *Journal* decided to disaggregate:** Alan Karben, "News You Can Reuse: Content Repurposing at *The Wall Street Journal Interactive Edition,*" *Markup Languages: Theory and Practice,* 1.1 (1999), http://mitpress.mit.edu/journals/MLANG/karben.pdf. This article describes an earlier system; the complete system I discuss in the book was brought online a few years later. Amusingly, at least twice during the first months after the switchover, I saw an XML code that slipped erroneously into the print edition—the modern equivalent of seeing *etaoin shrdlu.*

130 **it can add a brand new way to publish its content:** Talk by Alan Karben, "XML: Publishing on the Web Just Got Easier," Proceedings of Seybold San Franciso '97, 1997, http://seminars.seyboldreports.com/1997_san_francisco/10.html. Karben specifically comments on how stock market data displayed on the page is based on the content of the news story.

137 **No sane person will run an Internet connection to their refrigerator:** Paul Haas, "Paul's (Extra) Refrigerator," http://hamjudo.com/cgi-bin/refrigerator. This Web page is famous as one of the first that displayed the output of real-time

sensors. At one time, you could also use a Web page to wave to Paul's cats—one of the first Web pages that connected the Internet to actuators.

137 **efforts by local governments to legislate taxes, regulations, and "public safety" rules for IT service:** Charles M. Davidson, "VoIP," FCC Forum—December 1, 2003, http://www.fcc.gov/voip/presentations/davidson.ppt. At the time of this writing, the U.S. Federal Communications Commission continues to debate about Internet telephony services and how they would provide the actual physical location of callers during emergency telephone calls. These proposed rules, as well as the rules that require that cell phones provide detailed physical location information, allow government agencies to track the physical location of all telecommunication users.

Index

References to Endnotes are indicated as n.

About Disaggregate

Disaggregate provides consulting to help people understand, create, and apply revolutionary technology.

Services include:

- Project consulting, including new product development
- Long-term project assistance
- Training seminars
- Technology education

Disaggregate can be reached on the web at http://www.Disaggregate.com, via e-mail at speech@pobox.com, or by telephone at (+1) 773.764.8727.

About the Author

The author and family on vacation.
Top row: Yehuda, Moshe, Channan; Bottom row: Dreamer, Shorty, Comet

Moshe Yudkowsky received his Ph.D. in condensed matter physics from Northwestern University. After two years as assistant director of the Nuclear Magnetic Resonance Laboratory at the Department of Physics at Northwestern University, Moshe joined Bell Laboratories as a member of the technical staff. Moshe worked on a variety of projects at Bell Labs, including the first national deployment of AT&T's "robotic operator." After leaving AT&T, Moshe joined Dialogic Corporation as a senior system architect. Moshe worked with speech technology companies to find ways to simplify their transition to Dialogic hardware.

Moshe left Dialogic to found Disaggregate, a technology consulting firm that helps clients understand, create, and apply revolutionary technology.

In parallel with his work at AT&T and at Dialogic, Moshe chaired the Automatic Speech Recognition Task Force of the Enterprise Computer Telephony Forum, an industry trade group. In his final year with the forum, he served as technical chair of the organization.

Moshe is a founder of the Midwest Speech Technology Association, a professional organization centered in the Midwestern United States, and serves as chair of the organization. He is a member of the board of Applied Voice Input/Output Society, an international professional organization.

About Berrett-Koehler Publishers

Berrett-Koehler is an independent publisher dedicated to an ambitious mission: Creating a World that Works for All.

We believe that to truly create a better world, action is needed at all levels—individual, organizational, and societal. At the individual level, our publications help people align their lives and work with their deepest values. At the organizational level, our publications promote progressive leadership and management practices, socially responsible approaches to business, and humane and effective organizations. At the societal level, our publications advance social and economic justice, shared prosperity, sustainable development, and new solutions to national and global issues.

We publish groundbreaking books focused on each of these levels. To further advance our commitment to positive change at the societal level, we have recently expanded our line of books in this area and are calling this expanded line "BK Currents."

A major theme of our publications is "Opening Up New Space." They challenge conventional thinking, introduce new points of view, and offer new alternatives for change. Their common quest is changing the underlying beliefs, mindsets, institutions, and structures that keep generating the same cycles of problems, no matter who our leaders are or what improvement programs we adopt.

We strive to practice what we preach—to operate our publishing company in line with the ideas in our books. At the core of our approach is stewardship, which we define as a deep sense of responsibility to administer the company for the benefit of all of our "stakeholder" groups: authors, customers, employees, investors, service providers, and the communities and environment around us. We seek to establish a partnering relationship with each stakeholder that is open, equitable, and collaborative.

We are gratified that thousands of readers, authors, and other friends of the company consider themselves to be part of the "BK Community." We hope that you, too, will join our community and connect with us through the ways described on our website at www.bkconnection.com.

Be Connected

Visit Our Website

Go to www.bkconnection.com to read exclusive previews and excerpts of new books, find detailed information on all Berrett-Koehler titles and authors, browse subject-area libraries of books, and get special discounts.

Subscribe to Our Free E-Newsletter

Be the first to hear about new publications, special discount offers, exclusive articles, news about bestsellers, and more! Get on the list for our free e-newsletter by going to www.bkconnection.com.

Participate in the Discussion

To see what others are saying about our books and post your own thoughts, check out our blogs at www.bkblogs.com.

Get Quantity Discounts

Berrett-Koehler books are available at quantity discounts for orders of ten or more copies. Please call us toll-free at (800) 929-2929 or email us at bkp.orders@aidcvt.com.

Host a Reading Group

For tips on how to form and carry on a book reading group in your workplace or community, see our website at www.bkconnection.com.

Join the BK Community

Thousands of readers of our books have become part of the "BK Community" by participating in events featuring our authors, reviewing draft manuscripts of forthcoming books, spreading the word about their favorite books, and supporting our publishing program in other ways. If you would like to join the BK Community, please contact us at bkcommunity@bkpub.com.